AN INTRODUCTION TO
MATRIX STRUCTURAL ANALYSIS
AND FINITE ELEMENT METHODS

AN INTRODUCTION TO
MATRIX STRUCTURAL ANALYSIS
AND FINITE ELEMENT METHODS

Jean H. Prévost

Professor of Civil and Environmental Engineering
Princeton University

Serguei Bagrianski

Assistant in Instruction
Princeton University

World Scientific

NEW JERSEY · LONDON · SINGAPORE · BEIJING · SHANGHAI · HONG KONG · TAIPEI · CHENNAI · TOKYO

Published by

World Scientific Publishing Co. Pte. Ltd.

5 Toh Tuck Link, Singapore 596224

USA office: 27 Warren Street, Suite 401-402, Hackensack, NJ 07601

UK office: 57 Shelton Street, Covent Garden, London WC2H 9HE

Library of Congress Cataloging-in-Publication Data
Names: Prevost, Jean H. | Bagrianski, Serguei.
Title: An introduction to matrix structural analysis and finite element methods /
    Jean H. Prevost, Princeton University, Serguei Bagrianski, Princeton University.
Description: New Jersey : World Scientific, 2017. | Includes index.
Identifiers: LCCN 2016047610| ISBN 9789813206779 (hardcover) | ISBN 9789813206786 (softcover)
Subjects: LCSH: Structural analysis (Engineering) | Finite element method.
Classification: LCC TA645 .P7115 2017 | DDC 624.1/70151825--dc23
LC record available at https://lccn.loc.gov/2016047610

British Library Cataloguing-in-Publication Data
A catalogue record for this book is available from the British Library.

Printed in Singapore

Preface

This book is based on a course taught at Princeton University over the past 15 years to advanced undergraduate and graduate students in the civil and mechanical engineering departments. The semester-long course consisted of weekly lectures, precepts, and problem sets, as well as written exams and a final project. The original support materials were limited to overview lecture slides and sample code; over time, precept and course notes were also developed. This book combines these resources in a cohesive, stand-alone document accessible to anyone with a basic understanding of statics, calculus, linear algebra, and coding.

Matrix Structural Analysis (MSA) and Finite Element Methods (FEM) are typically classified as related but decoupled fields; this volume is unique in presenting both subjects in a cohesive framework. In our investigation of MSA, we derive formulations for truss, beam, and frame elements, which we use to develop the overarching framework of matrix analysis. The FEM chapters build on this foundation to develop numerical approximation techniques for solving boundary value problems in steady-state heat and linear elasticity. Diligently focused on coding, this book guides the reader from first principles to implementable algorithms.

This intensive, code-centric approach prepares the student or practitioner to critically assess the performance of commercial analysis packages and explore advanced literature on the subject. The reader should engage actively with the material by attempting to reproduce the presented code. Every algorithm is provided as complete MATLAB® code, an object-oriented language which may also be treated as pseudocode. New elements are validated using examples presented with complete code and comprehensive results.

Acknowledgements

We owe numerous people sincere gratitude for their contribution to the course and the book it catalyzed. First and foremost, we are indebted to Allison B. Halpern and Fabien Georget, who have served as dedicated instructors for the course and have provided their invaluable input to this book. Allison developed considerable portions of the original precept notes, contributed to examples presented in this text, and diligently edited the final manuscript. Fabien acted as a reliable advisor during the writing process and provided keen review of the book's technical content. We thank David Luet for helping to develop a predecessor to the code presented in this text. Our editors at WSPC, Steven Patt and Rochelle Kronzek, have been very supportive throughout the preparation of this document. We must also acknowledge our students, whose enthusiasm for the material and earnest feedback shaped the pedagogical strategy behind this book. We reserve the greatest gratitude for our families who have provided their unwavering support for us in this endeavor as they have for countless others: to my wife Carol and my children - JHP; to my parents Anna & Andrei and my brother Misha - SB.

JHP & SB

Contents

Chapter 1

Setting Up

Matrix Structural Analysis (MSA) and **Finite Element Methods (FEM)** are numerical analysis techniques which rely on the reduction of complex physical problems into sets of linear equations solved using computer algorithms. Prior to the 20th century, numerical analysis could only be executed by human or analog computers, which were slow, expensive, and prone to error. The advent of digital computers in the 1940s offered the possibility of accurately performing calculations that could be reduced to a set of simple tasks. Established matrix techniques, which could be reduced to entry-by-entry manipulation, proved to be particularly well-suited for computer execution. The emergence of the **Direct Stiffness Method (DSM)**, the predecessor of MSA and FEM, in the early 1950s facilitated the use of matrix techniques for the analysis of structural systems.

Commercialization of computers has resulted in the proliferation of MSA and FEM in academia and industry. The incredible improvement in the capacity of analysis packages has far outpaced user awareness of the underlying theory and inherent limitations of these tools. Defaults and user-friendly interfaces have made complex code increasingly more accessible to a broader user base while simultaneously obscuring the inner workings of black-box software packages. Although it may be argued that an engineer does not need to know how a software package works as long as it does, the engineer still bears the responsibility of correctly using the appropriate tool for a specific task. With the plethora of elements and analysis techniques available in modern software packages, the permutations of misuse far outnumber those of correct use. Like any method of analysis, MSA and FEM have a particular scope of application and best practice guidelines that cannot be relegated to settings in a program.

MSA and FEM, while grounded in scientific theory and mathematics, both come to life as code. Learning about these matrix techniques can follow one of two strategies: the invested reader can set out to read texts on the subject passively, learning the terminology and theory, even testing out the methods with an associated software package; or, he or she can code in the basic MSA/FEM program directly. While the former method facilitates a broader survey of the material, it is only through writing code that a novice can attain an acute and reliable understanding of MSA/FEM. Many of the characteristics and limitations of advanced software packages originate in basic code; only through an intimate understanding of the fundamental building blocks of MSA and FEM can the reader expect to appreciate the significance of higher level programs.

This text can be seen as a narrated recipe book for coding MSA and FEM; it presents a simple, but robust set of algorithms that make up a basic, but powerful analysis package. The reader is encouraged to engage actively with the code, implementing the various algorithms as they are presented. The content is tailored to an audience of undergraduates, graduate students, and practitioners in the fields of structural and mechanical engineering. Hence, the reader is assumed to have a basic understanding of statics, linear algebra, calculus, and coding. Derivations are designed to lead the reader from fundamental physical and mathematical theory directly to implementable code. Upon reading this text and testing out the code, the reader should feel prepared to read higher level texts and critically assess the results of commercially available analysis packages.

1.1 Scope

MSA and FEM originated as techniques for solving static and dynamic problems in mechanical and structural engineering but have since spread to a broader scope of application. This text is limited to problems in static linear elasticity and steady-state heat.

Trusses, beams, and **frames** delimit the traditional territory of MSA. For each element, we present the basic stiffness formulations and demonstrate how to extract internal element forces (axial force, moment, shear, and torsion as applicable).

We begin our investigation of FEM via **steady-state heat**. Besides being a common engineering problem, heat only has one degree of freedom (temperature) allowing for a simple derivation of the basic stiffness formulation. Next, we derive basic three- and four-node **elastic** elements and conclude with bending-capable **plates** and **shells**.

	Element		Dimensions			Degrees of Freedom						
	Type	Nodes	x	y	z	u	v	w	θ_x	θ_y	θ_z	T
MSA	Truss	2	✓			✓						
			✓	✓		✓	✓					
			✓	✓	✓	✓	✓	✓				
	Beam	2	✓			✓					✓	
			✓	✓				✓	✓	✓		
	Frame	2	✓			✓	✓				✓	
			✓	✓		✓	✓				✓	
			✓	✓	✓	✓	✓	✓	✓	✓	✓	
FEM	Heat	2	✓									✓
		3, 4	✓	✓								✓
	Elasticity	3, 4	✓	✓		✓	✓					
			✓	✓	✓	✓	✓	✓				
	Plate	4	✓	✓				✓	✓	✓		
	Shell	4	✓	✓		✓	✓	✓	✓	✓	✓	
			✓	✓	✓	✓	✓	✓	✓	✓	✓	

It is worth stressing that this book is concerned with **analysis**, not **design**. Though they may share variables and vocabularies, the tools of design and analysis are distinct. Analysis is broadly defined as the extraction of information that describes the performance of a particular physical system. Typically, the unknown information in a structural problem consists of reaction forces, internal element forces/stresses/strains, and displacements. This information may in turn inform design decisions regarding global geometry, element sections, and material choices. Of course, we cannot perform an analysis without a starting point for these variables; the exchange between analysis and design is thus iterative, in some cases permitting parametric optimization. The types of analysis covered in this book are far from the only methods available to the student or practicing engineer. Engineers should always verify any computer-aided analysis using a simplified set of hand calculations. Engineers designed highly efficient, elegant structural systems well before MSA and FEM came into existence. While new analytical techniques help fine-tune the final design, the importance of conceptual design via traditional techniques should not be overlooked.

It is important to moderate the accuracy of any analysis in reference to the reliability of the measurements or assumptions used. For instance, standard gravity on Earth varies from 9.83 m/s^2 at the North Pole to 9.78 m/s^2 at the equator, resulting in a variation of 0.5% in gravity across the world. Since most structural analysis is based on assumed standard gravity, reporting results to more than four significant figures is misleading. When we consider how much less reliable our predictions are of other forms of loading, material properties, support conditions, or fabrication accuracy, we can see the false security of reporting overly accurate analysis results.

1.1.1   A Basic Structural Analysis Problem

Structural analysis requires the engineer to formulate a real-world scenario as an **idealization** appropriate to a form of analysis. The engineering student typically forfeits this step to permit a pedagogical direction particular to the analysis undertaken. Although we do not divert from this strategy, the reader should remain cognizant of the idealizations inherent to any form of analysis.

In this section, we will analyze a basic truss structure with properties, supports, and loads as shown:

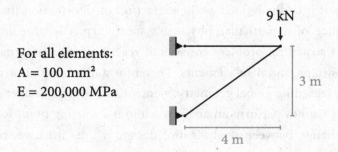

For all elements:
A = 100 mm²
E = 200,000 MPa

9 kN

3 m

4 m

Figure 1.1.   A basic structural problem composed of truss elements.

Our first step in solving this structural problem is to identify **knowns** and **unknowns**. The problem setup provides known element properties (sectional and material), geometry, applied loads, and support conditions. The unknowns consist of the support reactions, element forces/stresses/strains, and nodal displacements. To initiate our analysis, we label the nodes and elements before identifying the unknown nodal forces and displacements:

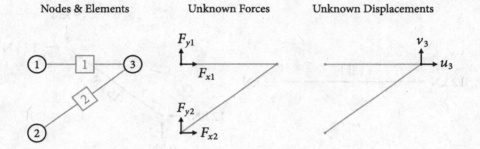

Figure 1.2. Labels and unknown forces & displacements.

Our first objective will be to solve for the unknown reactions using **equilibrium**. In two dimensions, there are three conditions of equilibrium:

$$\sum F_x = 0; \quad \sum F_y = 0; \quad \sum M = 0 \tag{1.1}$$

We have four unknowns but only three equations of equilibrium. By inspection, we note that the vertical reaction at the first node must be zero, $F_{y1} = 0$, since a truss element cannot transfer loads orthogonal to its axis. We can then use the three equilibrium equations to solve for the remaining unknown reactions, starting with horizontal equilibrium:

$$\sum F_x = 0 = F_{x1} + F_{x2}; \quad F_{x1} = -F_{x2} \tag{1.2}$$

Next, we employ vertical equilibrium:

$$\sum F_y = 0 = F_{y1} + F_{y2} - 9\,\text{kN}; \quad F_{y2} = 9\,\text{kN} \tag{1.3}$$

Then, we take moment equilibrium about node one:

$$\sum M_1 = 0 = (4\,\text{m})(-9\,\text{kN}) + (3\,\text{m})(F_{x2}); \quad F_{x2} = 12\,\text{kN} \tag{1.4}$$

We can now state our complete set of the recovered reaction forces:

$$F_{x1} = -12\,\text{kN}; \quad F_{y1} = 0\,\text{kN}; \quad F_{x2} = 12\,\text{kN}; \quad F_{y2} = 9\,\text{kN} \tag{1.5}$$

By drawing out the forces acting on each element, we can solve for the internal element forces directly:

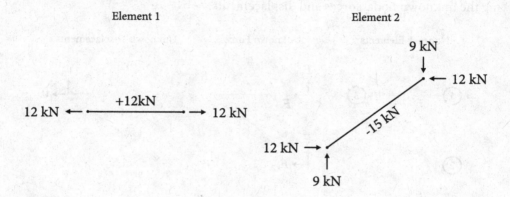

Figure 1.3.   Internal element forces.

We note that element 1 is subject to a tensile force designated with a positive sign; element 2 is in compression and thus marked with a negative sign. Throughout this book, forces, stresses, and strains are labelled following this convention; a positive sign is used to indicate elongation (an increase in length) while a negative sign indicates shortening (a decrease in length).

Having obtained the internal element forces, we find the axial stresses using the element cross-sectional areas, A:

$$\sigma_1 = \frac{F_1}{A} = \frac{+12\,\text{kN}}{100\,\text{mm}^2} = +120\,\text{MPa}; \quad \sigma_2 = -150\,\text{MPa} \tag{1.6}$$

We then use Young's modulus, E, to calculate the element strains:

$$\varepsilon_1 = \frac{\sigma_1}{E} = \frac{+120\,\text{MPa}}{200000\,\text{MPa}} = +0.0006; \quad \varepsilon_2 = -0.00075 \tag{1.7}$$

To find the nodal displacements, we can use any of several known techniques, such as Virtual Work or Force Method, but since we already know the element strains, we can calculate the displacements directly. The deformed length, L_f, of the two elements can be found by rearranging the equation for strain in terms of the initial undeformed length, L_i:

$$\varepsilon = \frac{L_f - L_i}{L_i}; \quad L_f = (1+\varepsilon)L_i \qquad (1.8)$$

The two deformed element lengths are then calculated to excessive accuracy:

$$L_1 = 4.0024 \, \text{m}; \quad L_2 = 4.99625 \, \text{m} \qquad (1.9)$$

With some trigonometry, which we will not show here, we find the horizontal and vertical displacements at the third node:

$$u_3 = 2.40 \, \text{mm}; \quad v_3 = -9.45 \, \text{mm} \qquad (1.10)$$

To complete our solution, we present the **Free Body Diagram (FBD)** and the **deformed shape**.

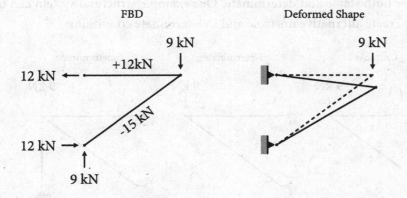

Figure 1.4. Complete solution presented as FBD and deformed shape.

These two diagrams follow a very specific representational strategy. For the FBD, supports are removed and replaced by the full set of forces, including applied loads and support reactions. If the analysis indicates that a force was opposite to the original assumed direction, we reverse the vector and write in the magnitude as a positive value; it is incorrect to label vectors with negative values. Null vectors should be omitted. Additionally, we label the bar forces with a positive or negative sign to indicate whether the element is in tension or compression. For the deformed shape, the support conditions and the applied loads are still drawn.

The deformed shape is typically exaggerated and plotted using a solid line. The undeformed shape is underlaid in a dotted line. If space permits, labelling of displacements is also a good idea.

Although these strategies may appear to be "just" questions of presentation and thus irrelevant to the "real" engineering work, the authors stress that clear presentation of a solution is critically important to any analysis. These two final diagrams give a clear and comprehensive summary of the analysis results. Because the FBD is labelled with correctly-oriented vectors, we can immediately verify that our structure is in equilibrium; the undeformed shape shows the tip of the truss deflecting not only downwards, but also to the right.

1.1.2 Questions of Determinacy

Our approach for solving the preceding problem only works for structural systems that are both stable and determinate. Our example structural system can be modified to create alternative unstable and indeterminate conditions:

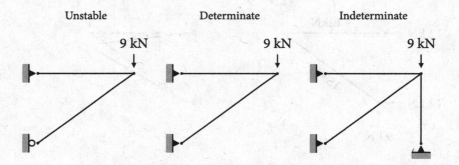

Figure 1.5. Levels of determinacy: unstable, determinate, and indeterminate.

If we release the vertical restraint in the bottom pin, turning it into a vertical roller, the structure becomes **unstable**. Though we could apply the same procedure we just demonstrated for the stable structure, our calculations would reveal that no combination of reaction forces could result in an equilibrated structure. This outcome is a limitation in the structural problem, not the analysis method employed. In this case, the instability is obvious. With more complex problems, it may be harder to detect instabilities, but the problem will remain unsolvable regardless of the analysis method used. Thus, we must always be careful to ensure that a structure is stable prior to performing any analysis.

An **indeterminate** or **redundant** structural system is one where all reactions and element forces cannot be solved using just equilibrium conditions. We can generate an indeterminate system by introducing a pinned vertical element to the example structure. By inspection, we note that four reaction forces are unknown, but we are still only equipped with three equations of equilibrium. Because we cannot even solve for the reactions, our system is considered **externally indeterminate**; if we could solve for the reactions, but not for the element forces, the system would be considered **internally indeterminate**. The behavior of an indeterminate structural system is influenced by the stiffness of its constituent elements. By significantly reducing the stiffness of elements in the system, the behavior of the system is drastically affected:

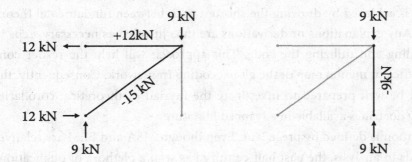

Figure 1.6. Effects of reduced element stiffness. Negligible stiffness indicated with gray shading.

When the element stiffnesses differ less dramatically, solving the structural problem requires more than just equilibrium conditions; we also need to employ kinematic and constitutive relationships. Several traditional techniques such as Virtual Work, Force Method, and Moment Distribution allow the partial analysis of certain categories of indeterminate structures, but only MSA and FEM integrate equilibrium, kinematic, and constitutive relationships in a truly general approach.

1.2 Presentation

This book has been prepared with the philosophy that presentation is indistinguishable from content. To this end, the three governing principles behind the design of this book are correctness, clarity, and convention.

Correctness is achieved by prioritizing functionality over theory; since we are developing a series of algorithms, it is more important to demonstrate that each procedure works correctly rather than explaining why. We still maintain a strong emphasis on derivation, but recognize that there are multiple paths to the same correct formulation. For instance, the truss element which we will first present in MSA can also be derived as an element of FEM using either the Variational Principle or the Method of Weighted Residual. We do not claim allegiance to any one method of derivation, but instead focus on demonstrating the aptitude of the presented techniques.

Clarity is achieved by drawing the shortest path between fundamental theory and code. Any explanations or derivations are thus justified as necessary steps in understanding and utilizing the code. This approach will help the reader construct an efficient mental map of the global coding framework. Consequently, the reader will be well prepared to investigate the myriad of theoretical corollaries and coding doctrines available in advanced literature.

Convention is defined by precedent. Even though MSA and FEM are relatively new fields in analysis, the past half-century has seen a plethora of publications on these subjects from which a reasonably stable vocabulary has emerged. Since most publications have been concerned with the theoretical aspects of MSA and FEM rather than the practicalities of implementing these theories in the code, we have a much more consistent collection of theoretical symbols and terms than coding syntax and structure. Consequently, the authors have established code variables by using terms equivalent to their theoretical counterparts and commonly accepted code syntax. When no precedent is available, new variable names are established according to nomenclature rules implied by precedent.

1.2.1 Graphic Conventions

Visual representation is integral to the material covered in this book. Sign convention for coordinates, vectors, and rotations follows the right-hand rule (thumb points right, index finger points up, and middle finger points out of page). 2D geometry is represented using traditional orthographic projection:

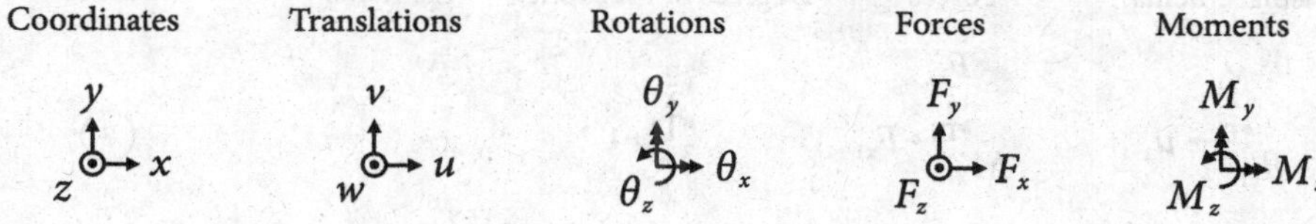

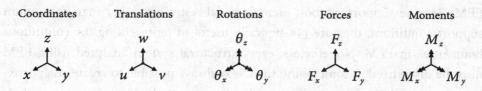

Figure 1.7. Sign conventions for orthographic projection.

3D geometry is presented using isometric projection:

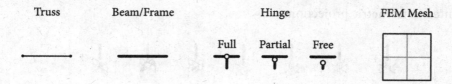

Figure 1.8. Sign conventions for isometric projection.

Elements are also presented using a consistent visual convention. Trusses cannot transmit moment, and thus are designated with a finer line than beams or frames. Hinges are indicated using a small empty circle positioned as to indicate the level of release. FEM meshes for heat, elasticity, plates, and shells, are all drawn with a dark line demarcating the bounding area/volume and lighter lines identifying the mesh.

Truss	Beam/Frame	Hinge			FEM Mesh
		Full	Partial	Free	

Figure 1.9. Element representation.

A variety of labelling schemes are used throughout the text. Displacements and forces are labelled using additional subscripts linked to the node. Degrees of freedom are labelled using numbers in place of the displacements/forces. Elements are identified using a square inscribed with the element number; nodes are identified using a circle inscribed with the node number.

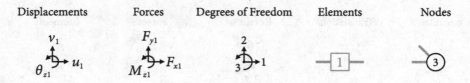

Figure 1.10. Labelling conventions.

Support symbols are used to indicate levels of restraint at nodes. These symbols are used for all elements in MSA and for discretized structural problems in FEM. Because support symbols identify nodal restraints rather than continuum support conditions, they are not typically useful in representing the continuous boundaries in FEM. Nonetheless, every structural system analyzed using FEM must be discretized at some point; thus, it is always possible to create an approximation of the boundary conditions using these symbols. These support symbols are defined in 2D orthographic projection:

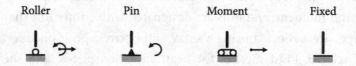

Figure 1.11. Support types in 2D with unrestrained displacements identified.

Permuted in three dimensions, we obtain a new set of restraints which can be presented in isometric projection:

Figure 1.12. Support types in 3D.

1.2.2 Matrices

The techniques studied in this book rely heavily on linear algebra, specifically for manipulating sets of linear equations. While the reader is assumed to have familiarity with the subject, we review some basics and make clear the conventions used in this text. Consider a simple example set of equations:

$$y_1 = C_{11}x_1 + C_{12}x_2$$
$$y_2 = C_{21}x_1 + C_{22}x_2 \tag{1.11}$$

These equations can be easily represented in matrix form:

$$\begin{Bmatrix} y_1 \\ y_2 \end{Bmatrix} = \begin{bmatrix} C_{11} & C_{12} \\ C_{21} & C_{22} \end{bmatrix} \begin{Bmatrix} x_1 \\ x_2 \end{Bmatrix} \tag{1.12}$$

The matrix equation can be more concisely represented as follows:

$$\{y\} = [C]\{x\} \tag{1.13}$$

The constants are stored in a 2×2 **matrix**, while the dependent and independent variables are stored in 2×1 **vectors**:

$$\{y\}_{2\times1} = \begin{Bmatrix} y_1 \\ y_2 \end{Bmatrix}; \quad [C]_{2\times2} = \begin{bmatrix} C_{11} & C_{12} \\ C_{21} & C_{22} \end{bmatrix}; \quad \{x\}_{2\times1} = \begin{Bmatrix} x_1 \\ x_2 \end{Bmatrix} \tag{1.14}$$

With this example, we note that matrices are denoted using square brackets, [ ], while vectors are denoted using curly brackets, { }. The sizes and indices of matrices are presented in order by **rows** and **columns**. For instance, a 3×2 matrix has three rows and two columns, while C_{12} identifies the entry in the first row and second column of the $[C]$ matrix.

Matrices permit a variety of useful operations including multiplication, inversion, and transposition. To help us demonstrate these operations, we define a new matrix, $[B]$:

$$[B]_{3\times2} = \begin{bmatrix} B_{11} & B_{12} \\ B_{21} & B_{22} \\ B_{31} & B_{32} \end{bmatrix} \tag{1.15}$$

Matrix **multiplication** is not commutative; the product of two matrices can only be calculated if the interior dimensions match. For instance, we can perform the multiplication $[B]_{3\times2}\,[C]_{2\times2}$, but not $[C]_{2\times2}\,[B]_{3\times2}$. The size of the product of two matrices is defined by the two outer dimensions. The product defined as $[A]=[B][C]$ will be 3×2. The ij entry of the product is equal to the dot product of the ith row of the first matrix and the jth column of the second matrix. The product of these two matrices will contain the entries:

$$A_{ij} = \sum_k B_{ik}C_{kj} \tag{1.16}$$

Summation iterates through k for the number of columns in $[B]$ or the number of rows in $[C]$ (these two numbers must be equal). The product may be defined explicitly:

$$[A] = \begin{bmatrix} B_{11} & B_{12} \\ B_{21} & B_{22} \\ B_{31} & B_{32} \end{bmatrix} \begin{bmatrix} C_{11} & C_{12} \\ C_{21} & C_{22} \end{bmatrix} = \begin{bmatrix} B_{11}C_{11}+B_{12}C_{21} & B_{11}C_{12}+B_{12}C_{22} \\ B_{21}C_{11}+B_{22}C_{21} & B_{21}C_{12}+B_{22}C_{22} \\ B_{31}C_{11}+B_{32}C_{21} & B_{31}C_{12}+B_{32}C_{22} \end{bmatrix} \tag{1.17}$$

It is worth noting that even though $[A]$ is the same size as $[B]$, the two matrices are very unlikely to share the same entries. There is a specific type of matrix called the **identity matrix**, $[I]$, which, when multiplied by another matrix, produces the matrix itself:

$$[B]=[B][I] \tag{1.18}$$

As the matrix equivalent to the number 1, the identity matrix is composed of ones along the diagonal entries and zeroes elsewhere:

$$[I] = \begin{bmatrix} 1 & 0 & \cdots & 0 \\ 0 & 1 & \cdots & 0 \\ \vdots & \vdots & \ddots & \vdots \\ 0 & 0 & \cdots & 1 \end{bmatrix} \tag{1.19}$$

The identity matrix is thus a **square matrix** (a matrix having the same number of rows and columns) whose dimensions are defined by its context. For instance, the size of the identity matrix will change depending on its position in the multiplication:

$$[B]_{3\times2} = [B]_{3\times2}[I]_{2\times2} = [I]_{3\times3}[B]_{3\times2} \qquad (1.20)$$

The identity matrix allows us to define the **inverse**, $[C]^{-1}$, of a matrix, $[C]$:

$$[C][C]^{-1} = [I] \qquad (1.21)$$

The inverse of a matrix parallels the reciprocal, $1/c$, of a scalar, c:

$$(c)(c)^{-1} = (c)(1/c) = 1 \qquad (1.22)$$

Just as the reciprocal only exists if the scalar c is nonzero, the inverse only exists if the matrix $[C]$ is **nonsingular**. A nonsingular (or invertible) matrix is a square matrix whose rows (and columns) are linearly independent. By definition, the **determinant** of a nonsingular matrix is nonzero. For a 2×2 matrix, the determinant is defined as follows:

$$|C| = C_{11}C_{22} - C_{12}C_{21} \qquad (1.23)$$

This expression allows us to define the inverse explicitly:

$$[C]^{-1} = \frac{1}{|C|}\begin{bmatrix} +C_{22} & -C_{12} \\ -C_{21} & +C_{11} \end{bmatrix} \qquad (1.24)$$

For larger matrices, finding a general algebraic form of the matrix inverse is rarely worthwhile. Instead, we can obtain the matrix inverse numerically using techniques such as Gaussian elimination or LU decomposition. When the matrix inverse is not required directly, but is instead a part of matrix expression, there are efficient techniques for obtaining the solution without solving the inverse.

Another important matrix operation is the **transpose**, which is obtained by swapping the columns and rows of a matrix:

$$[B]^T = \begin{bmatrix} B_{11} & B_{12} \\ B_{21} & B_{22} \\ B_{31} & B_{32} \end{bmatrix}^T = \begin{bmatrix} B_{11} & B_{21} & B_{31} \\ B_{12} & B_{22} & B_{32} \end{bmatrix} \qquad (1.25)$$

A matrix whose transpose is equal to itself is called a **symmetric** matrix:

$$[C]^T = [C] \tag{1.26}$$

A matrix whose inverse is equal to its transpose is called an **orthogonal** matrix:

$$[C]^T = [C]^{-1} \tag{1.27}$$

1.2.3 Code

In this book, code is presented using MATLAB® syntax. True to its abbreviation **MATrix LABoratory, MATLAB** is a high-level, object-oriented programming language particularly well-equipped for handling matrices. Readers more comfortable with another language are encouraged to use the provided code as pseudocode.

While a proper introduction to MATLAB is outside the scope of this book, we will introduce some fundamentals in this section. Through this text, we present code mainly as **functions** stored in individual .m files. An example function `multMatrices` (stored as "multMatrices.m") is presented below:

```
1 function [A] = multMatrices(B,C)                  % function header
2
3 rowB = size(B,1); colB = size(B,2);        % extract matrix dimensions
4 rowC = size(C,1); colC = size(C,2);
5 rowA = rowB;      colA = colC;
6
7 if colB == rowC          % ensure that inner matrix dimensions agree
8   A = zeros(rowA,colA);                        % initialize output A
9   for i = 1:rowA
10     for j = 1:colA
11       for k = 1:rowC
12         A(i,j) = A(i,j) + B(i,k)*C(k,j);
13       end
14     end
15   end
16 end
```

This function is not particularly useful; all it does is find the product of the matrix multiplication of B and C and stores it in A. This function is, however, useful in illustrating a good range of MATLAB functionality.

The first line of code is the function header, which defines the name of the function (this name has to match the file name), the inputs (identified in the parentheses to the immediate right of the function name), and the outputs (identified in the square brackets to the left of the equal sign). The label function must be used to differentiate a **function** (which can take inputs and generate outputs) from a **script** (which simply executes a block of instructions). Though all the functions we use in this text are stored in their own .m files, it is also possible to create local functions accessible only in the file where they are stored. A function or script may be called directly in the command window or in another routine.

Because MATLAB does not require variables to be initialized or data types to be specified, we need to be particularly careful in tracking variable assignments. Multiple assignments on the same line are permitted as long as a semicolon (suppresses output) or comma (prints new line to command line) is used to separate the individual assignments. On lines 3 through 5, we initialize and assign six new scalars (two per line): rowB, colB, rowC, colC, rowA, and colA.

The two inputs in our function are assumed to be two-dimensional matrices. We find the number of rows or columns they contain by using the size(matrix,dims) function. To initialize the empty output matrix, A, we use the zeros(rows,cols) function. We access individual entries in a matrix by identifying the row and column index in the parentheses respectively, i.e. A(i,j).

To make our code perform non-trivial operations, we need to use conditional and iterative blocks. The basic **if…end** statement is used in our example function to determine if the inner dimensions of the two input matrices are equal. We can also use **if…else/elseif…end** statements if we have multiple conditions to consider. When several choices are available, a **switch…case…end** set of operations is recommended. We can perform iterations using either **for** or **while** statements, but we use **for** loops in this text exclusively. The counter is initialized in the first statement, **for** i = startIndex:endIndex, and incremented automatically by one at the completion of each loop. We note that both types of blocks are terminated by **end** statements rather than parentheses. To make presentation of the code as clear as possible, key words in blocks are bolded and hierarchical indentation is maintained.

We can see from our example code that three nested loops are used to iterate through the rows of matrix A, the columns of matrix A, and the columns of matrix B (also rows of matrix C). Matrix A is filled entry by entry. Fortunately, MATLAB has built-in matrix functionality that can automatically multiply matrices together; the whole function can be replaced by the operation, A = B*C. In fact, MATLAB already has functions that perform all of the matrix operations discussed in the previous section. The inverse of a nonsingular square matrix is found using inv(C). If we are interested in finding the product of a matrix and the inverse of another matrix, we can avoid the computational burden of solving for the inverse by using B = A/C or B = C\A. To find the transpose of a matrix, we simply append an apostrophe after the matrix, B'.

To make functions compact, we also use cells and strings throughout our main code. **Cells** store variably sized matrices under one name. For instance, the outputs of our **runAnalysis** function are stored in two cell variables, output and process, to reduce the variable clutter of such a large operation. We also use **strings** in the **runAnalysis** function to generate function handles for the different element stiffness functions to avoid changing the hardcoding of the program. While these techniques are particular to MATLAB, similar strategies may be found in other programming languages.

There are many tricks and simplifications that an adept user of MATLAB may employ to simplify and condense code. The struggle lies in balancing brevity with comprehension. As much as possible, variable naming, code structure, and tricks used throughout this book are designed to make the code as comprehensible to the reader as possible.

Chapter 2

Truss Element

In Chapter 1, we looked at an example structural problem for which we were able to find the support reactions and element forces using only equilibrium conditions. Because an equilibrium-based analysis can only be used to solve determinate systems, its scope of application is significantly limited. In order to analyze indeterminate structures, it is necessary to incorporate kinematic and constitutive relationships.

A stiffness-based analysis requires that constitutive, kinematic, and equilibrium conditions are satisfied in every part of a structure. To set up this analysis, we need to first deconstruct the structural system into its constituent elements. For each type of element (truss, beam, frame, etc...), it is necessary to formulate general relationships between deformation and loading. We assemble these element-specific stiffness relationships into a set of global relationships describing the behavior of the global structural system. In reassembling the structural system, we introduce additional unknowns (nodal displacements and forces), but we also add more information (equations). The laws of mechanics guarantee that as long as the global structure is stable, we will have sufficient equations to solve for all unknown reactions and displacements.

In this chapter, we find the stiffness of a simple truss element first in its local, 1D orientation and then in two and three dimensions. Using this general formulation of a truss stiffness element, we solve the problem from Chapter 1. We conclude by implementing the truss element as code.

2.1 Deriving the Truss Element Stiffness Equation

A structure subject to loading will deform. The magnitude of deformation is determined by the structure's **stiffness**, which is informed by the material properties, sectional geometry, orientation, and connectivity of its constituent parts. In this section, we derive the truss element stiffness matrix based on basic constitutive, kinematic, and equilibrium relationships.

2.1.1 Constitutive Relationship

In structural mechanics, a **constitutive** relationship is defined as a material property relating strain to stress. Robert Hooke first identified the relationship in 1678 via the Latin anagram, *Ut tension sic vis*, which translates *as the extension, so the force*. While Hooke's Law established proportionality between load and deformation, it was Leonhard Euler who quantified the linear relationship in 1727:

$$\sigma = E\varepsilon \qquad (2.1)$$

The **elastic** or **Young's modulus**, E, represents the slope on the stress-strain curve for which the linear relationship holds. If we look at the stress-strain curve for common structural materials such as steel and concrete, we observe that a linear assumption covers a significant fraction of the material capacity. When we consider that safety factors commonly used in design provide at least a 25% buffer to the ultimate capacity, we can be typically comfortable with the assumption of linear-elastic behavior in the design of structural systems.

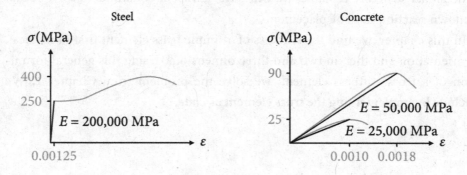

Figure 2.1. Typical stress-strain curves for steel and concrete.

For structural steel, a strain of 0.00125 with a corresponding yield stress of 250 MPa marks the commonly accepted limit of elastic behavior. The elastic modulus for steel is defined as the tangent of the elastic portion of the stress-strain curve:

$$E_{steel} = \frac{250\,\text{MPa}}{0.00125} = 200,000\,\text{MPa} \tag{2.2}$$

Structural concrete is less consistent, ranging in strength from 25 MPa to 90 MPa with corresponding strains of 0.0010 to 0.0018 for compression. The elastic modulus for concrete typically uses the secant interpolated to the breaking stress:

$$E_{concrete} = \frac{25\,\text{MPa}}{0.0010} \; to \; \frac{90\,\text{MPa}}{0.0018} = 25,000\,\text{MPa} \; to \; 50,000\,\text{MPa} \tag{2.3}$$

Since all of the structural elements studied in this book are based on linear elasticity, we must assess our results to ensure not only that stresses remain within the material limits, but also that strains remain within the limits of elasticity. More advanced techniques are available for non-linear and/or inelastic behaviors, but these techniques are outside the scope of this book.

2.1.2 Kinematic Relationship

A **kinematic** relationship relates local strains to global deformations. If we define a local x-axis to be directly aligned with a truss element, we can express the **normal strain** as the derivative of **axial deformation**, u:

$$\varepsilon = \frac{du}{dx} \tag{2.4}$$

A truss element subject only to forces applied to its ends will experience deformation that varies linearly over the element length:

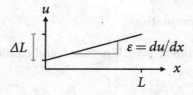

Figure 2.2. Strain in truss subject to nodal forces.

The normal strain for the truss can thus be described simply as the total change in length, ΔL, divided by the undeformed length of the element, L:

$$\varepsilon = \frac{\Delta L}{L} \tag{2.5}$$

2.1.3 Equilibrium Relationship

To equilibrate the truss, we must apply equal and opposite forces; outward forces put the truss element into tension, generating positive internal stress and strains:

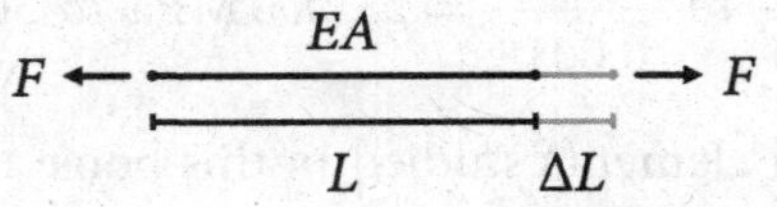

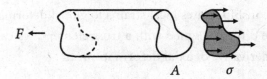

Figure 2.3.   Equilibrium of a truss element subject to axial forces.

The only relevant equilibrium condition (horizontal) is trivially satisfied:

$$\sum F_x = 0 = (-F) + F \tag{2.6}$$

Taking a sectional cut we equilibrate the external force with the internal stresses:

Figure 2.4.   Equilibrating external and internal forces in a truss.

If we apply equilibrium again, we find the more useful relationship:

$$F = A\sigma \tag{2.7}$$

2.1.4 Stiffness Matrix Formulation

We have now established constitutive, kinematic, and equilibrium equations:

$$\sigma = E\varepsilon; \quad \varepsilon = \frac{\Delta L}{L}; \quad F = A\sigma \tag{2.8}$$

Combining these three relationships allows us to arrive at the stiffness equation for a truss element directly relating deformation, ΔL, to applied force, F:

$$F = A\sigma = A(E\varepsilon) = AE\left(\frac{\Delta L}{L}\right) = \frac{EA}{L}\Delta L \tag{2.9}$$

In order to analyze more complex structures composed of truss elements connected at nodes, we need to express this stiffness relationship using the nodal displacements and forces. We start by labelling the two nodes of element e:

Figure 2.5. Node numbering for element e.

In this text, we will use superscripts to and subscripts to denote element numbering and nodal numbering or dofs respectively. The element deformation can be expressed as the difference between the two nodal displacements:

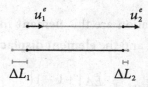

Figure 2.6. Nodal displacements.

Nodal displacements adhere to global sign convention; a positive displacement always translates the node to the right. A positive change in length corresponds to a negative first node displacement and a positive second node displacement:

$$\Delta L = \Delta L_1 + \Delta L_2 = -u_1^e + u_2^e \tag{2.10}$$

Next, we identify the nodal forces:

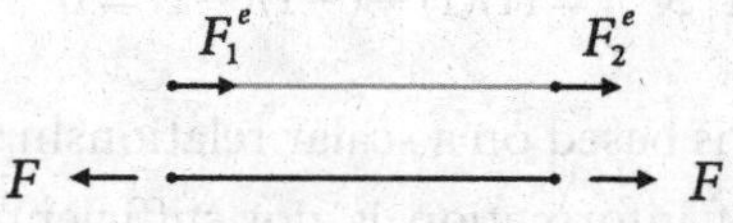

$$F_1^e \qquad F_2^e$$

$$F \leftarrow \qquad \longrightarrow F$$

Figure 2.7. Nodal forces.

Unlike the displacements, the nodal forces are not cumulative, but instead equal and opposite. A positive force (tension) occurs when the first nodal force is negative and the second nodal force is positive:

$$F = -F_1^e = F_2^e \tag{2.11}$$

With these two relationships, we express the stiffness equation as follows:

$$F = -F_1^e = F_2^e = \frac{EA}{L}\Delta L = \frac{EA}{L}\left(-u_1^e + u_2^e\right) \tag{2.12}$$

More conveniently, we can express the two equations in matrix form:

$$\begin{Bmatrix} F_1^e \\ F_2^e \end{Bmatrix} = \frac{EA}{L}\begin{bmatrix} +1 & -1 \\ -1 & +1 \end{bmatrix}\begin{Bmatrix} u_1^e \\ u_2^e \end{Bmatrix} \tag{2.13}$$

For ease of representation, the vectors and matrices can be reduced to:

$$\left\{ F^e \right\} = \left[K^e \right]\left\{ d^e \right\} \tag{2.14}$$

The components of this equation are the **element force vector**, $\left\{ F^e \right\}$, the **element stiffness matrix**, $\left[K^e \right]$, and the **element displacement vector**, $\left\{ d^e \right\}$:

$$\left\{ F^e \right\} = \begin{Bmatrix} F_1^e \\ F_2^e \end{Bmatrix}; \quad \left[K^e \right] = \frac{EA}{L}\begin{bmatrix} +1 & -1 \\ -1 & +1 \end{bmatrix}; \quad \left\{ d^e \right\} = \begin{Bmatrix} u_1^e \\ u_2^e \end{Bmatrix} \tag{2.15}$$

The element stiffness matrix demonstrates two properties common to all stiffness matrices studied in this book. First, the stiffness matrix is **symmetric**:

$$\left[K^e \right] = \left[K^e \right]^T; \quad K_{ij}^e = K_{ji}^e \tag{2.16}$$

Second, the stiffness matrix is **singular**:

$$\det\left[K^e \right] = (1)(1) - (-1)(-1) = 0 \tag{2.17}$$

Since our stiffness matrix is based on a scalar relationship, the singularity is inevitable; the total amount of information is not sufficient to generate two unique equations.

2.1.5 Rotating the Element

If we have a structural system composed entirely of parallel truss elements, the 2×2 truss element formulation might be sufficient, but trussed structures are generally composed of variably inclined struts and ties. A structural system is designated in a global coordinate system out of elements that can be defined in either the local or global coordinate system. Thus, we need to extend the stiffness equation to a general Cartesian formulation. We begin in two dimensions, with an element rotated counterclockwise from the global x-axis by an angle, φ:

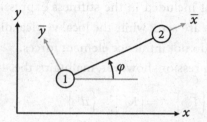

Figure 2.8. Inclined truss element.

The **global (xy) coordinate system** can be defined arbitrarily (but typically aligns with the page), while the **local ($\overline{xy}$) coordinate system** uses the first node as the origin and aligns the local x-axis to the truss axis. To be explicit, we identify the nodal displacements and forces in both coordinate systems:

Local Coordinate System   Global Coordinate System

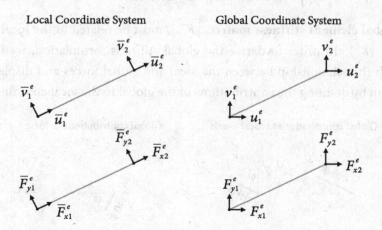

Figure 2.9. Local and global coordinate systems for 2D truss element.

The local stiffness equation maintains its 2×2 local element stiffness matrix:

$$\left\{ \overline{F}^e \right\}_{2\times 1} = \left[k^e \right]_{2\times 2} \left\{ \overline{d}^e \right\}_{2\times 1} \tag{2.18}$$

Note that local forces and displacements are designated using a macron, $^{-}$, while the local stiffness is identified using the lowercase, k:

$$\left\{ \overline{F}^e \right\}_{2\times 1} = \left\{ \begin{array}{c} \overline{F}_{x1}^e \\ \overline{F}_{x2}^e \end{array} \right\}; \quad \left[k^e \right]_{2\times 2} = \frac{EA}{L} \begin{bmatrix} +1 & -1 \\ -1 & +1 \end{bmatrix}; \quad \left\{ \overline{d}^e \right\}_{2\times 1} = \left\{ \begin{array}{c} \overline{u}_1^e \\ \overline{u}_2^e \end{array} \right\} \tag{2.19}$$

Local vertical dofs are not included in the stiffness expression; the local vertical nodal forces, $\overline{F}_{y1}^e$ and $\overline{F}_{y2}^e$, are zero, while the local vertical displacements, $\overline{v}_1^e$ and $\overline{v}_2^e$, may be nonzero, but do not influence element forces.

The global stiffness expression, however, maintains the 4×4 formulation:

$$\left\{ F^e \right\}_{4\times 1} = \left[K^e \right]_{4\times 4} \left\{ d^e \right\}_{4\times 1} \tag{2.20}$$

For completeness, we define the global components explicitly:

$$\left\{ F^e \right\} = \left\{ \begin{array}{c} F_{x1}^e \\ F_{y1}^e \\ F_{x2}^e \\ F_{y2}^e \end{array} \right\}; \quad \left[K^e \right] = \begin{bmatrix} K_{11}^e & K_{12}^e & K_{13}^e & K_{14}^e \\ K_{21}^e & K_{22}^e & K_{23}^e & K_{24}^e \\ K_{31}^e & K_{32}^e & K_{33}^e & K_{34}^e \\ K_{41}^e & K_{42}^e & K_{43}^e & K_{44}^e \end{bmatrix}; \quad \left\{ d^e \right\} = \left\{ \begin{array}{c} u_1^e \\ v_1^e \\ u_2^e \\ v_2^e \end{array} \right\} \tag{2.21}$$

The **global element stiffness matrix**, $\left[K^e \right]$, must be related to the **local stiffness matrix**, $\left[k^e \right]$. In order to derive the global stiffness formulation, we must first establish the relationship between the local and global forces and displacements. We begin by defining the contributions of the global to the local coordinates:

Global contributions to local x-axis   Global contributions to local y-axis

Figure 2.10.   Contributions of global axes to local axes.

We express the coordinate transformation algebraically as follows:

$$\bar{x} = +x \cos \varphi + y \sin \varphi$$
$$\bar{y} = -x \sin \varphi + y \cos \varphi$$

(2.22)

The equations can be reformatted in matrix form using $s = \sin \varphi$ and $c = \cos \varphi$:

$$\begin{Bmatrix} \bar{x} \\ \bar{y} \end{Bmatrix} = \begin{bmatrix} +c & +s \\ -s & +c \end{bmatrix} \begin{Bmatrix} x \\ y \end{Bmatrix} = [Q] \begin{Bmatrix} x \\ y \end{Bmatrix}$$

(2.23)

- This formulation permits us to define the **rotation matrix**, $[Q]$. Unlike the element stiffness matrix, the rotation matrix is nonsingular. Hence, we can inverse this relationship to find the transformation from local to global coordinates:

$$\begin{Bmatrix} x \\ y \end{Bmatrix} = \begin{bmatrix} +c & +s \\ -s & +c \end{bmatrix}^{-1} \begin{Bmatrix} \bar{x} \\ \bar{y} \end{Bmatrix} = \frac{1}{(c)(c)-(s)(s)} \begin{bmatrix} +c & -s \\ +s & +c \end{bmatrix} \begin{Bmatrix} \bar{x} \\ \bar{y} \end{Bmatrix} = \begin{bmatrix} +c & -s \\ +s & +c \end{bmatrix} \begin{Bmatrix} \bar{x} \\ \bar{y} \end{Bmatrix}$$

(2.24)

It is important to note that while the rotation matrix is not symmetric, it is **orthogonal**, meaning that its transpose is equal to its inverse:

$$[Q]^T = [Q]^{-1}$$

(2.25)

The rotation matrix also defines the translation of forces and displacements from local to global coordinate systems:

$$\begin{Bmatrix} \bar{F}_x^e \\ \bar{F}_y^e \end{Bmatrix} = [Q] \begin{Bmatrix} F_x^e \\ F_y^e \end{Bmatrix}; \quad \begin{Bmatrix} \bar{u}^e \\ \bar{v}^e \end{Bmatrix} = [Q] \begin{Bmatrix} u^e \\ v^e \end{Bmatrix}$$

(2.26)

These two sets of relationships will help us derive the stiffness matrix in our global sign convention. We begin by inverting the force relationship:

$$\begin{Bmatrix} F_x^e \\ F_y^e \end{Bmatrix} = [Q]^T \begin{Bmatrix} \bar{F}_x^e \\ \bar{F}_y^e \end{Bmatrix}$$

(2.27)

Next, we need to expand this formulation to two nodes; recalling that the off-axis nodal forces, $\bar{F}_{y1}$ and $\bar{F}_{y2}$, are zero, we can directly relate the 4×1 global force vector to the 2×1 local force vector:

$$\begin{Bmatrix} F_{x1}^e \\ F_{y1}^e \\ F_{x2}^e \\ F_{y2}^e \end{Bmatrix} = \begin{bmatrix} +c & -s & 0 & 0 \\ +s & +c & 0 & 0 \\ 0 & 0 & +c & -s \\ 0 & 0 & +s & +c \end{bmatrix} \begin{Bmatrix} \overline{F}_{x1}^e \\ 0 \\ \overline{F}_{x2}^e \\ 0 \end{Bmatrix} = \begin{bmatrix} +c & 0 \\ +s & 0 \\ 0 & +c \\ 0 & +s \end{bmatrix} \begin{Bmatrix} \overline{F}_{x1}^e \\ \overline{F}_{x2}^e \end{Bmatrix} \tag{2.28}$$

The translation from local to global displacements is identically defined:

$$\begin{Bmatrix} \overline{u}_1^e \\ \overline{u}_2^e \end{Bmatrix} = \begin{bmatrix} +c & +s & 0 & 0 \\ 0 & 0 & +c & +s \end{bmatrix} \begin{Bmatrix} u_1^e \\ v_1^e \\ u_2^e \\ v_2^e \end{Bmatrix} \tag{2.29}$$

Plugging in equations (2.28) and (2.29) into (2.18) produces the relationship:

$$\begin{Bmatrix} F_{x1}^e \\ F_{y1}^e \\ F_{x2}^e \\ F_{y2}^e \end{Bmatrix} = \begin{bmatrix} +c & 0 \\ +s & 0 \\ 0 & +c \\ 0 & +s \end{bmatrix} \begin{bmatrix} k^e \end{bmatrix} \begin{bmatrix} +c & +s & 0 & 0 \\ 0 & 0 & +c & +s \end{bmatrix} \begin{Bmatrix} u_1^e \\ v_1^e \\ u_2^e \\ v_2^e \end{Bmatrix} \tag{2.30}$$

We define the **transformation matrix**, $\begin{bmatrix} T^e \end{bmatrix}$, as follows:

$$\begin{bmatrix} T^e \end{bmatrix} = \begin{bmatrix} +c & +s & 0 & 0 \\ 0 & 0 & +c & +s \end{bmatrix} \tag{2.31}$$

We can now define the global element stiffness matrix concisely:

$$\begin{bmatrix} K^e \end{bmatrix}_{4\times4} = \begin{bmatrix} T^e \end{bmatrix}_{4\times2}^T \begin{bmatrix} k^e \end{bmatrix}_{2\times2} \begin{bmatrix} T^e \end{bmatrix}_{2\times4} \tag{2.32}$$

With a bit of algebra, the global element stiffness matrix can be defined explicitly:

$$\begin{bmatrix} K^e \end{bmatrix}_{4\times4} = \frac{EA}{L} \begin{bmatrix} c^2 & sc & -c^2 & -sc \\ sc & s^2 & -sc & -s^2 \\ -c^2 & -sc & c^2 & sc \\ -sc & -s^2 & sc & s^2 \end{bmatrix} \tag{2.33}$$

2.1.6 General Formulation

The rotation matrix that we just developed was defined with reference to an angle of rotation. We can also express this rotation matrix using nodal coordinates:

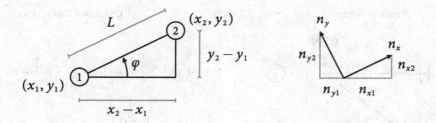

Figure 2.11. Inclined geometry and directional cosines.

The nodal coordinates can be used to find the **direction cosines**, n_x and n_y, which populate the rotation matrix:

$$[Q] = \begin{bmatrix} +c & +s \\ -s & +c \end{bmatrix} = \frac{1}{L} \begin{bmatrix} x_2 - x_1 & y_2 - y_1 \\ y_1 - y_2 & x_2 - x_1 \end{bmatrix} = \begin{bmatrix} n_{x1} & n_{x2} \\ n_{y1} & n_{y2} \end{bmatrix} \tag{2.34}$$

We can also reformulate the transformation matrix using the directional cosines:

$$\left[T^e \right] = \begin{bmatrix} n_{x1} & n_{x2} & 0 & 0 \\ 0 & 0 & n_{x1} & n_{x2} \end{bmatrix} \tag{2.35}$$

The general form of the transformation matrix for a truss element rotated into one, two, or three **spatial dimensions**, nsd, is defined as follows:

$$\left[T^e \right]_{2 \times (2 \times nsd)} = \begin{bmatrix} \{n_x\}_{1 \times nsd} & \{0\}_{1 \times nsd} \\ \{0\}_{1 \times nsd} & \{n_x\}_{1 \times nsd} \end{bmatrix} \tag{2.36}$$

We can use this formulation to establish the general expression for the global truss stiffness matrix:

$$\left[K^e \right]_{(2 \times nsd) \times (2 \times nsd)} = \left[T^e \right]^T_{(2 \times nsd) \times 2} \left[k^e \right]_{2 \times 2} \left[T^e \right]_{2 \times (2 \times nsd)} \tag{2.37}$$

2.2 Solving the Same Problem

In this section, we will demonstrate how to solve the example from Chapter 1 using a stiffness approach, instead of the equilibrium method. Recall the setup:

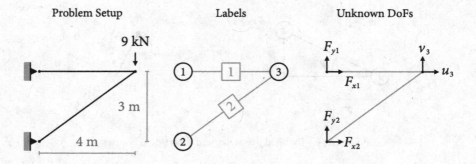

Figure 2.12. For all elements, E = 200,000 MPa and A = 100 mm².

The structure is composed of two elements, whose geometry and nodal dofs are defined as follows:

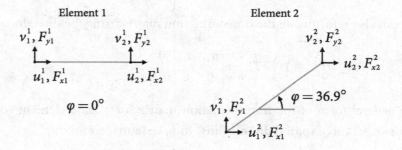

Figure 2.13. Element geometries and degrees of freedom.

We begin by calculating the two local stiffness matrices:

$$\left[k^1\right] = \frac{(200,000\,\text{MPa})(100\,\text{mm}^2)}{(4000\,\text{mm})}\begin{bmatrix} +1 & -1 \\ -1 & +1 \end{bmatrix} = \begin{bmatrix} +5000 & -5000 \\ -5000 & +5000 \end{bmatrix}\frac{\text{N}}{\text{mm}} \qquad (2.38)$$

$$\left[k^2\right] = \frac{(200,000\,\text{MPa})(100\,\text{mm}^2)}{(5000\,\text{mm})}\begin{bmatrix} +1 & -1 \\ -1 & +1 \end{bmatrix} = \begin{bmatrix} +4000 & -4000 \\ -4000 & +4000 \end{bmatrix}\frac{\text{N}}{\text{mm}} \qquad (2.39)$$

Next, we obtain the two transformation matrices:

$$\left[T^1\right] = \frac{1}{4\,m}\begin{bmatrix} 4\,m & 0 & 0 & 0 \\ 0 & 0 & 4\,m & 0 \end{bmatrix} = \begin{bmatrix} 1 & 0 & 0 & 0 \\ 0 & 0 & 1 & 0 \end{bmatrix} \tag{2.40}$$

$$\left[T^2\right] = \frac{1}{5\,m}\begin{bmatrix} 4m & 3m & 0 & 0 \\ 0 & 0 & 4m & 3m \end{bmatrix} = \begin{bmatrix} 0.8 & 0.6 & 0 & 0 \\ 0 & 0 & 0.8 & 0.6 \end{bmatrix} \tag{2.41}$$

It is now easy to obtain the global element stiffness matrices:

$$\left[K^1\right] = \left[T^1\right]^T\left[k^1\right]\left[T^1\right] = \begin{bmatrix} +5000 & 0 & -5000 & 0 \\ 0 & 0 & 0 & 0 \\ -5000 & 0 & +5000 & 0 \\ 0 & 0 & 0 & 0 \end{bmatrix} \frac{\text{N}}{\text{mm}} \tag{2.42}$$

$$\left[K^2\right] = \left[T^2\right]^T\left[k^2\right]\left[T^2\right] = \begin{bmatrix} +2560 & +1920 & -2560 & -1920 \\ +1920 & +1440 & -1920 & -1440 \\ -2560 & -1920 & +2560 & +1920 \\ -1920 & -1440 & +1920 & +1440 \end{bmatrix} \frac{\text{N}}{\text{mm}} \tag{2.43}$$

These two stiffness matrices represent a set of eight linear equations:

$$\begin{aligned}
F_{x1}^1 &= +5000u_1^1 - 5000u_2^1 \\
F_{y1}^1 &= 0 \\
F_{x2}^1 &= -5000u_1^1 + 5000u_2^1 \\
F_{y2}^1 &= 0 \\
F_{x1}^2 &= +2560u_1^2 + 1920v_1^2 - 2560u_2^2 - 1920v_2^2 \\
F_{y1}^2 &= +1920u_1^2 + 1440v_1^2 - 1920u_2^2 - 1440v_2^2 \\
F_{x2}^2 &= -2560u_1^2 - 1920v_1^2 + 2560u_2^2 + 1920v_2^2 \\
F_{y2}^2 &= -1920u_1^2 - 1440v_1^2 + 1920u_2^2 + 1440v_2^2
\end{aligned} \tag{2.44}$$

In total, there are only six nodal forces and six nodal displacements; since elements must remain connected, global and element displacements are shared:

$$\left\{ u_1 \ v_1 \ u_2 \ v_2 \ u_3 \ v_3 \right\} = \left\{ u_1^1 \ v_1^1 \ u_1^2 \ v_1^2 \ (u_2^1, u_2^2) \ (v_2^1, v_2^2) \right\} \tag{2.45}$$

Equilibrium dictates that element forces sum up to produce the global forces.

$$\left\{ F_{x1} \ F_{y1} \ F_{x2} \ F_{y2} \ F_{x3} \ F_{y3} \right\} = \left\{ F_{x1}^1 \ F_{y1}^1 \ F_{x1}^2 \ F_{y1}^2 \ F_{x2}^1 + F_{x2}^2 \ F_{y2}^1 + F_{y2}^2 \right\} \quad (2.46)$$

Introducing these global relationships, we can reduce our eight element specific equations to just six featuring only the global dofs:

$$\begin{aligned}
F_{x1} &= +5000u_1 - 5000u_3 \\
F_{y1} &= 0 \\
F_{x2} &= +2560u_2 + 1920v_2 - 2560u_3 - 1920v_3 \\
F_{y2} &= +1920u_2 + 1440v_2 - 1920u_3 - 1440v_3 \\
F_{x3} &= -5000u_1 - 2560u_2 - 1920v_2 + 7560u_3 + 1920v_3 \\
F_{y3} &= -1920u_2 - 1440v_2 + 1920u_3 + 1440v_3
\end{aligned} \quad (2.47)$$

The horizontal and vertical displacements at the first two notes are set to zero:

$$u_1 = v_1 = u_2 = v_2 = 0 \quad (2.48)$$

We also know the applied forces at the unrestrained nodes:

$$F_{x3} = 0; \quad F_{y3} = -9000 \,\text{N} \quad (2.49)$$

Plugging in these six values, we can solve for the unknown displacements:

$$u_3 = +2.40 \,\text{mm}; \quad v_3 = -9.45 \,\text{mm} \quad (2.50)$$

Likewise, we can easily find the reactions:

$$F_{x1} = -12 \,\text{kN}; \quad F_{y1} = 0; \quad F_{x2} = 12 \,\text{kN}; \quad F_{y2} = 9 \,\text{kN} \quad (2.51)$$

The internal element forces can be found by manipulating the global element stiffness matrix expression:

$$\left\{ \begin{matrix} -F \\ +F \end{matrix} \right\} = \left\{ \begin{matrix} F_1^e \\ F_2^e \end{matrix} \right\} = \left\{ \overline{F}^e \right\} = \left[k^e \right] \left[T^e \right] \left\{ d^e \right\} \quad (2.52)$$

For our problem, the local element forces are found to be:

$$\left\{ \overline{F}^1 \right\} = \left\{ \begin{matrix} -12 \\ +12 \end{matrix} \right\} \text{kN}; \quad \left\{ \overline{F}^2 \right\} = \left\{ \begin{matrix} +15 \\ -15 \end{matrix} \right\} \text{kN} \quad (2.53)$$

By plotting out the FBD and deformed shape, we confirm that these results exactly match those found using the equilibrium method:

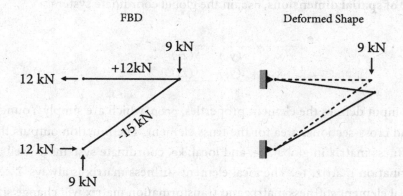

Figure 2.14. Summary of analysis results.

This example establishes the aptitude of a stiffness-based analysis; in the next chapter, we will systematize this approach to make it implementable as code.

2.3 Code Recipe

Using the mathematical formulation established in this chapter, we can write the function **Ke_truss** that generates the element stiffness matrix for a truss element.

```
1 function [Ke,ke,Te] = Ke_truss(xe,prop)
2
3 nsd = size(xe,2);                      % number of spatial dimensions
4 E = prop(1);                                    % Young's modulus
5 A = prop(2);                               % cross-sectional area
6
7 nx = xe(2,:) - xe(1,:);                      % orientation vector
8 L = norm(nx);                                      % truss length
9 nx = nx/L;                         % normalized orientation vector
10
11 % Initialize local stiffness, ke
12 ke = (E*A/L)*[+1,-1;-1,+1];
13
14 % Generate rotation matrix, Qe
15 Te = [nx zeros(1,nsd); zeros(1,nsd) nx];
16
17 % Find global element stiffness, Ke
18 Ke = Te'*ke*Te;
```

The first input defines the nodal coordinates, xe, storing each set of nodal coordinates in its individual row. The number of columns in this variable determines the **number of spatial dimensions**, nsd, in the global coordinate system.

xe	x	(y)	(z)
1	x_1	(y_1)	(z_1)
node 2	x_2	(y_2)	(z_2)

The second input defines the element properties, prop, which are simply Young's modulus and cross-sectional area for the truss element. The function outputs the element stiffness matrix in global, Ke, and local, ke, coordinate systems as well as the transformation matrix, Te. The local element stiffness matrix is always 2×2, but the global element stiffness matrix and transformation matrix will change size based on the number of spatial dimensions.

To demonstrate the performance of the truss stiffness function, we write a simple script to solve our example problem:

```
1 % exChpt2 - Analysis of a 2D truss in N and mm base units
2 clear;
3
4 % Define geometrical and material properties using N and mm
5 x1 = [0      0];                          % nodal coordinates, mm
6 x2 = [0 -3000];
7 x2 = [4000  0];
8 prop = [200000 100];         % modulus, MPa, & x-sectional area, mm^2
9
10 % Find element stiffness matrices
11 [K1,k1,T1] = Ke_truss([x1; x3],prop);
12 [K2,k2,T2] = Ke_truss([x2; x3],prop);
13
14 % Solve for unknown displacements
15 Kuu = K1(3:4,3:4) + K2(3:4,3:4);     % find constants in last two eqs
16 Pu = [0; 9000];                          % known applied loads, N
17 du = inv(Kuu)*Pu;            % solve for the last two displacements
18
19 % Solve for unknown reactions
20 Kus = [K1(1:2,3:4); K2(1:2,3:4)]; % find constants in first four eqs
21 Rs = Kus*du;                    % solve for the first four forces
22
23 % Solve for internal forces
24 F1 = k1*T1*[0; 0; du(1); du(2)];
25 F2 = k2*T2*[0; 0; du(1); du(2)];
```

In this example, we begin by defining the nodal coordinates, x1, x2, and x3, (using the first node as the origin) and the shared element properties, prop. We call on the element stiffness matrix function twice to generate the two sets of element stiffness and transformation matrices, [K1, k1, T1] and [K2, k2, T2].

To solve for the unknown displacements, we assemble the constants of the last two equations in (2.47) into a 2×2 stiffness matrix we call Kuu. By inverting this matrix and multiplying by a 2×1 vector, Pu, containing the known applied loads, we find the two unknown displacements, u_3 and v_3, stored in the 2×1 vector du.

Next, we assemble the constants for the first four equations relating to the nonzero displacements into a 4×2 matrix, Kus. Multiplying this matrix by du provides us with a 4×1 matrix, Rs, containing the reaction forces.

Finally, we find the 2×1 local element force vectors, F1 and F2, using 4×1 displacement vectors populated with the values stored in du. The results of our analysis are thus stored in four variables, du, Rs, F1, and F2:

du	(mm)		Rs	(kN)		F1	(kN)		F2	(kN)
u3	+2.40		Fx1	-12.0		F1	-12.0		F1	+15.0
v3	-9.45		Fy1	0		F2	+12.0		F2	-15.0
			Fx2	12.0						
			Fy2	9.0						

While the **Ke_truss** function is both general and robust, the example script we used is neither. Not only the inputs, but also the solution is fully hard-coded. In order to analyze any other structure, the entire script must be rewritten including the matrix operations to solve for the unknown displacements and forces. In the next chapter, we will develop a general approach for solving a structural system composed of multiple elements.

Chapter 3

Global Stiffness Equation

In Chapter 2, we used constitutive, kinematic, and equilibrium relationships to derive the matrix stiffness formulation for a truss element. We also demonstrated how to solve a simple structural problem using a stiffness-based approach. While the element stiffness formulation was general and robust, the method for analyzing the example structure was explicit and inflexible. In this chapter, we will formulate a systematic approach for the matrix analysis of structures; we will implement this method as code in Chapter 4. To demonstrate this procedure, we will use a slightly more complex example:

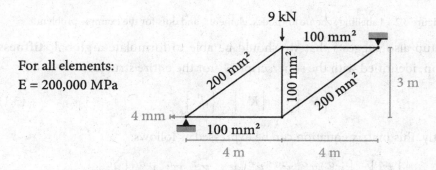

Figure 3.1. Indeterminate structural problem setup.

This problem has several specific qualities that highlight the capacity of a stiffness-based approach. First, this structure is indeterminate; neither the reactions nor the element forces can be solved only using equilibrium conditions. Second, the system is subject both to applied loads and nonzero prescribed displacements (achieved, for instance, through settlement). Third, the elements have variable properties (specifically, cross-sectional areas).

3.1   Assembling the Global Stiffness Equation

Our first step in formulating a general stiffness-based approach is to establish a clear labelling and identification system. We still identify nodes and elements, but we must now also label **degrees of freedom** (dofs), which enumerate the directional nodal forces and displacements. Conventionally, dofs are enumerated in order by node and direction. As the keen reader may have noticed, we always know either the force or displacement (but not both) at each dof; in fact, the number of dofs is equal to the total number of unknowns and the number of equations. For our example, a convenient labelling scheme is presented below:

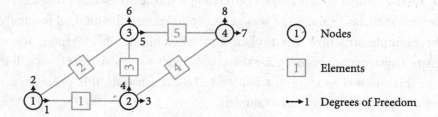

Figure 3.2.   Labelling scheme for nodes, elements, and dofs for the example problem.

This setup also suggests that we should be able to formulate a **global stiffness equation**, identified with the superscript G, for the entire structure:

$$\left\{F^G\right\}_{8\times1} = \left[K^G\right]_{8\times8}\left\{d^G\right\}_{8\times1} \tag{3.1}$$

Explicitly, this matrix equation can be expressed as follows:

$$\begin{Bmatrix} F_1^G \\ F_2^G \\ F_3^G \\ F_4^G \\ F_5^G \\ F_6^G \\ F_7^G \\ F_8^G \end{Bmatrix} = \begin{bmatrix} K_{11}^G & K_{12}^G & K_{13}^G & K_{14}^G & K_{15}^G & K_{16}^G & K_{17}^G & K_{18}^G \\ K_{21}^G & K_{22}^G & K_{23}^G & K_{24}^G & K_{25}^G & K_{26}^G & K_{27}^G & K_{28}^G \\ K_{31}^G & K_{32}^G & K_{33}^G & K_{34}^G & K_{35}^G & K_{36}^G & K_{37}^G & K_{38}^G \\ K_{41}^G & K_{42}^G & K_{43}^G & K_{44}^G & K_{45}^G & K_{46}^G & K_{47}^G & K_{48}^G \\ K_{51}^G & K_{52}^G & K_{53}^G & K_{54}^G & K_{55}^G & K_{56}^G & K_{57}^G & K_{58}^G \\ K_{61}^G & K_{62}^G & K_{63}^G & K_{64}^G & K_{65}^G & K_{66}^G & K_{67}^G & K_{68}^G \\ K_{71}^G & K_{72}^G & K_{73}^G & K_{74}^G & K_{75}^G & K_{76}^G & K_{77}^G & K_{78}^G \\ K_{81}^G & K_{82}^G & K_{83}^G & K_{84}^G & K_{85}^G & K_{86}^G & K_{87}^G & K_{88}^G \end{bmatrix} \begin{Bmatrix} d_1^G \\ d_2^G \\ d_3^G \\ d_4^G \\ d_5^G \\ d_6^G \\ d_7^G \\ d_8^G \end{Bmatrix} \tag{3.2}$$

More concisely, this set of equations can be represented as a summation:

$$F_P^G = \sum_Q K_{PQ}^G d_Q^G \quad \text{where} \quad P, Q \in [1 : neq] \tag{3.3}$$

Here, P indexes the rows of both the force vector and the stiffness matrix while Q indexes the columns of the stiffness matrix and the rows of the displacement vector. The variable, neq, defines the total **number of equations** in the system (eight for our example). It is worth noting that even though these indices reference dofs, they still represent directional components at each node. At a given node n, the directional forces can be expressed as $F_{xn} = F_{2n-1}$ and $F_{yn} = F_{2n}$, while the directional displacements can be expressed as $u_n = d_{2n-1}$ and $v_n = d_{2n}$.

The element stiffness equation may be presented using a similar summation:

$$F_p^e = \sum_q K_{pq}^e d_q^e \quad \text{where} \quad p, q \in [1 : nen] \tag{3.4}$$

The lowercase indices, p and q, represent **local indices,** while the uppercase indices, P and Q, represent **global indices**. The components of the element stiffness equation can be presented equivalently in either local or global indices:

$$F_p^e = F_P^e; \quad K_{pq}^e = K_{PQ}^e; \quad d_q^e = d_Q^e \tag{3.5}$$

When we transform an element stiffness matrix from local to global indices, we effectively map the local 4×4 element matrix to the full 8×8 global matrix. We demonstrate this operation visually for element 4 of the example structure:

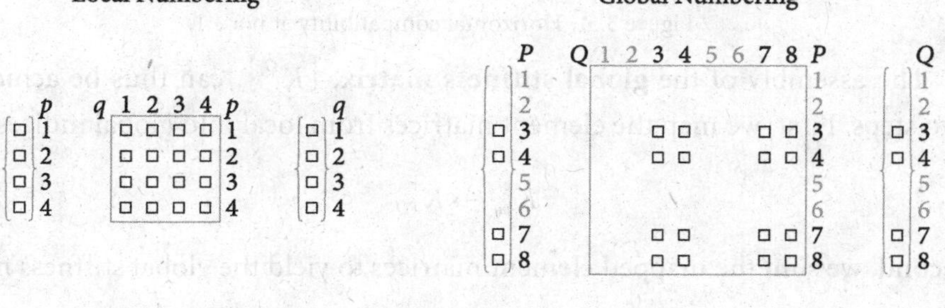

Figure 3.3. Local and global dof numbering for element 4.

Assembly of the global stiffness matrix relies on nodal equilibrium and element compatibility. **Equilibrium** necessitates that nodal element forces sum to produce the global force at each dof:

$$F_P^G = \sum F_P^e \tag{3.6}$$

We can demonstrate equilibrium at the first dof:

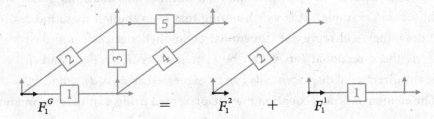

Figure 3.4. Horizontal equilibrium at node 1.

Compatibility requires that global nodes displace identically to element nodes:

$$d_Q^G = d_Q^e \tag{3.7}$$

We can similarly demonstrate compatibility at the first dof:

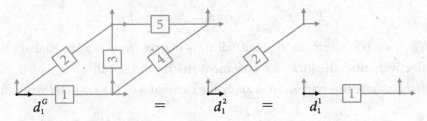

Figure 3.5. Horizontal compatibility at node 1.

The assembly of the **global stiffness matrix**, $[K^G]$, can thus be achieved in two steps. First, we map the element matrices from local into global indices:

$$K_{pq}^e \rightarrow K_{PQ}^e \tag{3.8}$$

Second, we sum the mapped element matrices to yield the global stiffness matrix:

$$K_{PQ}^G = \sum K_{PQ}^e \tag{3.9}$$

We demonstrate how the global stiffness matrix is populated visually:

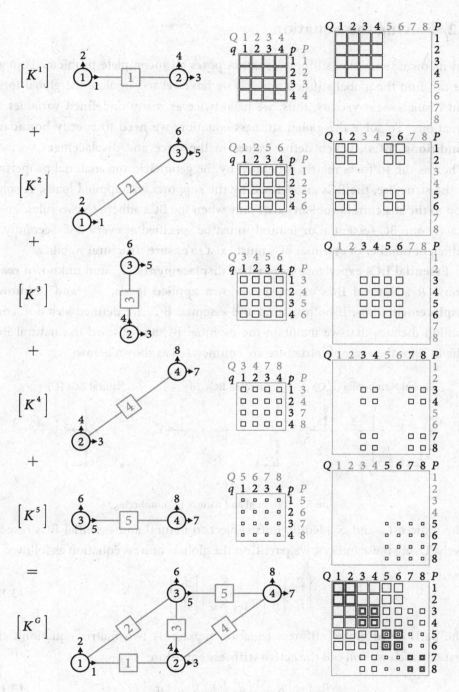

Figure 3.6. Global matrix assembly for example structure.

3.2　Solving the Equation

On its own, the global stiffness equation poses an incomplete problem. Though we can find the global stiffness matrix, we have yet to populate the global force and displacement vectors; thus, we have twice as many undefined variables as equations. To solve the global stiffness equation, we need to specify **boundary conditions (BCs)**, which define values in the force and displacement vectors. Whereas the stiffness matrix is defined by the geometric and material properties of the structure, the BCs are defined by the supports and applied loads. A solution to the structural problem exists only when the BCs adhere to two rules. First, exactly one BC (essential or natural) must be specified at every dof. Secondly, a sufficient number of essential BCs must exist to ensure structural stability.

Essential BCs experience **prescribed displacements**, d_s, and unknown **reactions**, R_s. **Natural BCs** experience known **applied loads**, P_U, and **unknown displacements**, d_U. If both natural and essential BCs are defined at a dof, convention dictates that we maintain the essential BC and discard the natural BC. The BCs for the example structure are enumerated as shown below:

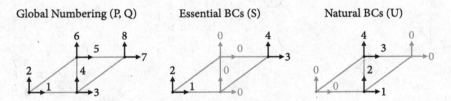

Global Numbering (P, Q)　　　Essential BCs (S)　　　Natural BCs (U)

Figure 3.7.　Essential and natural BC numbering.

The indices U and S identify dofs subject to natural and essential BCs respectively. Using these indices, we **partition** the global stiffness equation as follows:

$$\begin{Bmatrix} P_U \\ R_S \end{Bmatrix} = \begin{bmatrix} K_{UU} & K_{US} \\ K_{SU} & K_{SS} \end{bmatrix} \begin{Bmatrix} d_U \\ d_S \end{Bmatrix} \tag{3.10}$$

The partitioned global stiffness equation represents two matrix equations, the first of which we will call the **active stiffness equation**:

$$\{P_U\} = [K_{UU}]\{d_U\} + [K_{US}]\{d_S\} \tag{3.11}$$

The components of this equation include the **active stiffness matrix**, $[K_{UU}]$, and the **active force vector**, $\{P_U\}-[K_{US}]\{d_s\}$. The dofs associated with the natural BCs are commonly called the **active dofs**. In order to find the unknown displacement vector, $\{d_U\}$, we rearrange the equation:

$$\{d_U\}=[K_{UU}]^{-1}(\{P_U\}-[K_{US}]\{d_s\}) \tag{3.12}$$

Once we have found the unknown displacements, we use the **reaction stiffness equation** to calculate the reactions.

$$\{R_s\}=[K_{SU}]\{d_U\}+[K_{SS}]\{d_s\} \tag{3.13}$$

Since the only matrix that needs to be inverted is the active stiffness matrix, it is the only partition of the global stiffness matrix that must be explicitly assembled. We can extract the active partition of the global stiffness matrix by selecting entries that have columns and rows associated with natural dofs:

Figure 3.8. Extracting the active stiffness matrix from the global stiffness matrix.

We can also assemble the active stiffness matrix from element contributions:

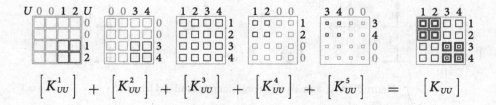

$$[K^1_{UU}] + [K^2_{UU}] + [K^3_{UU}] + [K^4_{UU}] + [K^5_{UU}] = [K_{UU}]$$

Figure 3.9. Assembling the active stiffness matrix from the element contributions.

The other stiffness partitions only appear in matrix products $[K_{US}]\{d_S\}$, $[K_{SU}]\{d_U\}$, and $[K_{SS}]\{d_S\}$ which can be calculated element by element:

$$[K]\{d\} = \sum_e [K^e]\{d^e\} \tag{3.14}$$

Although assembling the global stiffness matrix may appear like an intuitive and necessary step, the assembly and storage of a large matrix unnecessarily increases the computational burden. As a result, the efficient approach is to assemble only the active stiffness matrix and then perform all remaining matrix multiplications element by element.

3.3 Implementation

The general stiffness-based approach can be summarized in five steps, which we will describe and demonstrate in this section.

3.3.1 Label Nodes, Elements, and Degrees of Freedom

We begin by summarizing the labelling schemes we have already established:

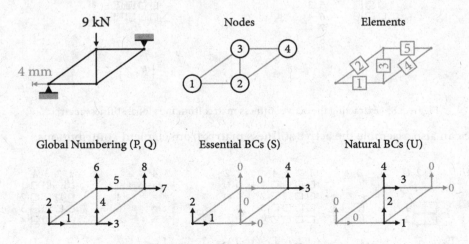

Figure 3.10. Summary of nodal, element, and dof labelling.

3.3.2 Generate Element Stiffness Matrices

Next, we generate the element stiffness matrices for the five truss elements. Since all of the elements are 2D trusses, the element matrices will be 4×4 in size.

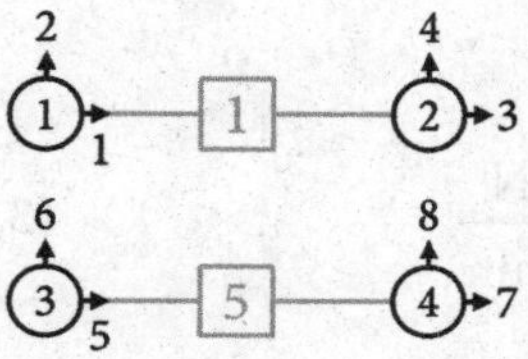

$$
\left[K^1 \right] = \left[K^5 \right] =
\begin{bmatrix}
5000 & 0 & -5000 & 0 \\
 & 0 & 0 & 0 \\
 & & 5000 & 0 \\
sym & & & 0
\end{bmatrix}
\frac{N}{mm}
$$

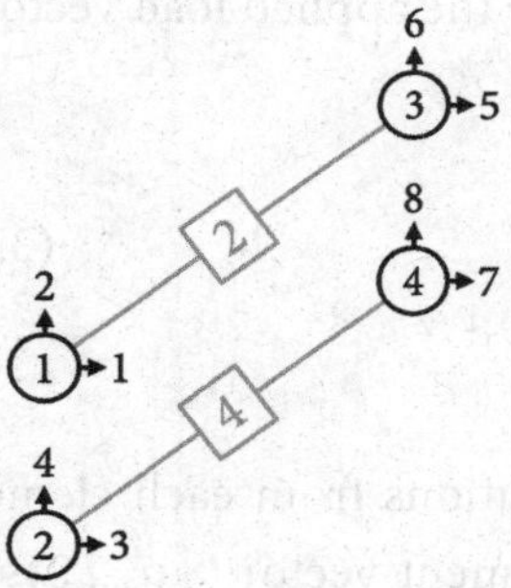

$$
\left[K^2 \right] = \left[K^4 \right] =
\begin{bmatrix}
5120 & 3840 & -5120 & -3840 \\
 & 2880 & -3840 & -2880 \\
 & & 5120 & 3840 \\
sym & & & 2880
\end{bmatrix}
\frac{N}{mm}
$$

$$
\left[K^3 \right] =
\begin{bmatrix}
0 & 0 & 0 & 0 \\
 & 6667 & 0 & -6667 \\
 & & 0 & 0 \\
sym & & & 6667
\end{bmatrix}
\frac{N}{mm}
$$

Figure 3.11. Element stiffness matrices.

Because elements 1 & 5 and 2 & 4 are identical in geometry and properties, their element stiffness matrices are also identical. Since element stiffness matrices are symmetrical, we only need to present the upper triangular portion of each matrix.

3.3.3   Assemble Active Stiffness Matrix and Active Force Vector

In order to find the unknown displacements, we first need to assemble the active stiffness matrix, $[K_{UU}]$, and the active force vector, $\{P_U\} - [K_{US}]\{d_s\}$. The active stiffness matrix is assembled directly from element contributions:

$$[K_{UU}] = \begin{bmatrix} 10120 & 3840 & 0 & 0 \\ & 9547 & 0 & 6667 \\ & & 10120 & 3840 \\ sym & & & 9547 \end{bmatrix} \frac{\text{N}}{\text{mm}} \tag{3.15}$$

In order to find the active force vector, we first populate the applied load vector:

$$\{P_U\} = \begin{Bmatrix} 0 \\ 0 \\ 0 \\ -9000 \end{Bmatrix} \text{N} \tag{3.16}$$

Then, we look at the prescribed displacement contributions from each element. For efficiency of computation, the prescribed displacement vector, $\{d_s\}$, is reduced to nonzero components (global dof 1):

$$\{d_s\} = \begin{Bmatrix} -4 \\ 0 \\ 0 \\ 0 \end{Bmatrix} \text{mm} = \{-4\} \text{mm} \tag{3.17}$$

The prescribed displacement contribution for element 1 is trivially found:

$$[K_{US}^1]\{d_s^1\} = \begin{bmatrix} -5000 \\ 0 \end{bmatrix} \{-4\} = \begin{Bmatrix} 20000 \\ 0 \end{Bmatrix} \text{N} \tag{3.18}$$

Since we only need to consider elements subject to nonzero prescribed displacements, the only remaining contribution is from element 2:

$$[K_{US}^2]\{d_s^2\} = \begin{bmatrix} -5120 \\ -3840 \end{bmatrix} \{-4\} = \begin{Bmatrix} 20480 \\ 15360 \end{Bmatrix} \text{N} \tag{3.19}$$

The total prescribed displacement force vector, $[K_{US}]\{d_s\}$, is summed from these two element contributions:

$$[K_{US}]\{d_s\} = \sum_e [K_{US}^e]\{d_s^e\} = \begin{Bmatrix} 20000 \\ 0 \\ 0 \\ 0 \end{Bmatrix} + \begin{Bmatrix} 0 \\ 0 \\ 20480 \\ 15360 \end{Bmatrix} = \begin{Bmatrix} 20000 \\ 0 \\ 20480 \\ 15360 \end{Bmatrix} \text{N} \qquad (3.20)$$

We now have all of the components for the active force vector:

$$\{P_U\} - [K_{US}]\{d_s\} = \begin{Bmatrix} 0 \\ 0 \\ 0 \\ -9000 \end{Bmatrix} - \begin{Bmatrix} 20000 \\ 0 \\ 20480 \\ 15360 \end{Bmatrix} = \begin{Bmatrix} -20000 \\ 0 \\ -20480 \\ -24360 \end{Bmatrix} \text{N} \qquad (3.21)$$

3.3.4 Solve for Unknown Displacements

To solve for the unknown displacements, we use the active stiffness equation, plugging in the active stiffness matrix and active force vector that we just found:

$$\{d_U\} = \begin{bmatrix} 10120 & 3840 & 0 & 0 \\ & 9547 & 0 & 6667 \\ & & 10120 & 3840 \\ sym & & & 9547 \end{bmatrix}^{-1} \begin{Bmatrix} -20000 \\ 0 \\ -20480 \\ -24360 \end{Bmatrix} = \begin{Bmatrix} -1.087 \\ -2.343 \\ -0.513 \\ -3.982 \end{Bmatrix} \text{mm} \qquad (3.22)$$

Upon calculating the unknown displacements, we draw the deformed shape:

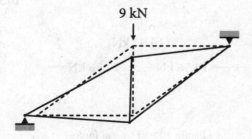

Figure 3.12. Deformed shape.

Although equation (3.22) implies that we need to find the inverse of the active stiffness matrix, we can more efficiently solve for the unknown displacements using Gaussian elimination and avoid calculating the matrix inverse explicitly.

3.3.5 Post-Process to Find Element Forces and Reactions

The most efficient method for post-processing is to calculate the element forces before assembling the reaction force vector. For each element, we do not need to distinguish between the prescribed and previously unknown displacements; since we now know all of the displacements, we can use them indiscriminately. The force vector in global coordinates for element 1 is found as follows:

$$\left\{F^1\right\} = \left[K^1\right]\left\{d^1\right\} = \begin{bmatrix} 5000 & 0 & -5000 & 0 \\ & 0 & 0 & 0 \\ & & 5000 & 0 \\ sym & & & 0 \end{bmatrix} \begin{Bmatrix} -4.000 \\ 0 \\ -1.087 \\ -2.343 \end{Bmatrix} = \begin{Bmatrix} -14.56 \\ 0 \\ +14.56 \\ 0 \end{Bmatrix} \text{kN} \qquad (3.23)$$

By repeating this calculation for the remaining elements, we generate the complete set of element forces, which we document in diagram form:

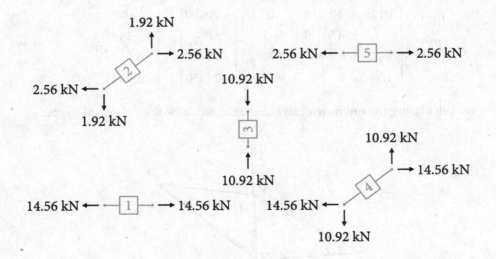

Figure 3.13. Element forces.

By summing the element force contributions at nodes, we find the full set of global forces acting on the structure, simultaneously determining the reactions and verifying the applied loads. These results are summarized using the FBD:

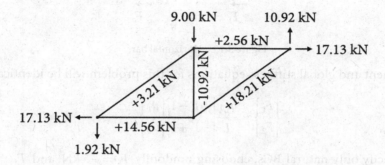

Figure 3.14. Free Body Diagram (FBD) with element forces.

Convention dictates we must also label the FBD with bar forces, which we obtain following the method established in the previous chapter. We demonstrate this calculation for the first element:

$$\left\{\overline{F}^1\right\} = \left[k^1\right]\left[T^1\right]\left\{d^1\right\} = \left[k^1\right]\left[T^1\right] \begin{Bmatrix} -4.000 \\ 0 \\ -1.087 \\ -2.343 \end{Bmatrix} = \begin{Bmatrix} -14.56 \\ +14.56 \end{Bmatrix} \text{kN} \qquad (3.24)$$

3.4 Instability and Singularity

Although the general method just demonstrated can be applied to any structural problem, we will only arrive at a solution if a sufficient set of essential BCs are prescribed. All of the element stiffness matrices we study in this text will be singular; since the global stiffness matrix is entirely assembled from element stiffness matrices, it will also be singular. The active stiffness matrix only becomes nonsingular when a sufficient number of linearly dependent columns and rows are removed from the global stiffness matrix. Hence, an insufficient number of essential BCs will result in a singular active stiffness matrix and an unstable structure.

To demonstrate instability and singularity, consider a simple 1D structure composed of a single truss element:

Figure 3.15.   Horizontal bar.

The element and global stiffness equations for this problem will be identical:

$$\begin{Bmatrix} F_1 \\ F_2 \end{Bmatrix} = \frac{EA}{L} \begin{bmatrix} +1 & -1 \\ -1 & +1 \end{bmatrix} \begin{Bmatrix} u_1 \\ u_2 \end{Bmatrix} \tag{3.25}$$

If we specify only natural BCs, choosing randomly $F_1 = -5\,\text{kN}$ and $F_2 = 10\,\text{kN}$, we end up with a system that is not in equilibrium:

Figure 3.16.   Unequilibrated forces.

If we define the stiffness, $EA/L = 2000\,\text{N/mm}$, and introduce the relationship, $\Delta L = u_2 - u_1$, we observe that there is no set of displacements that will produce an equilibrated set of applied forces:

$$\begin{aligned} -5000 &= 2000\left(u_1 - u_2 \right) \rightarrow \Delta L = 2.5\,\text{mm} \\ 10000 &= 2000\left(u_2 - u_1 \right) \rightarrow \Delta L = 5.0\,\text{mm} \end{aligned} \tag{3.26}$$

We might assume that we can resolve this issue by specifying the applied forces more strategically, for instance using $F_1 = -F_2 = -5\,\text{kN}$:

Figure 3.17.   Equilibrated forces.

This arrangement now provides us with a consistent deformation:

$$\begin{aligned} -5000 &= 2000\left(u_1 - u_2 \right) \rightarrow \Delta L = 2.5\,\text{mm} \\ +5000 &= 2000\left(u_2 - u_1 \right) \rightarrow \Delta L = 2.5\,\text{mm} \end{aligned} \tag{3.27}$$

However, we do not have enough information to solve for the nodal displacements; any number of paired displacements can produce the same elongation:

$$\Delta L = 2.5 = u_2 - u_1 = 2.5 - 0 = -2.5 - (-5.0) = 5000 - 4997.5 \qquad (3.28)$$

Whereas an unbalanced set of forces does not allow a single solution of displacements, a balanced set of forces produces an infinite set of solutions.

An insufficient number of essential BCs will lead to one of three forms of structural instability: a **global instability**, where no set of reactions can equilibrate the system, an **internal instability**, where the internal element forces cannot be equilibrated, and a **passive instability**, where instability may be triggered by a set of loads different from those currently applied.

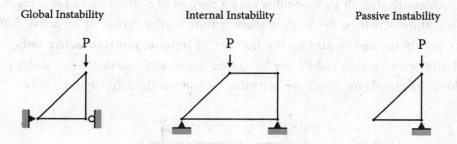

Figure 3.18. Forms of structural instability.

Mathematically, any of these types of structural instability will produce a **singular** active stiffness matrix. For the simple horizontal bar that we have been investigating in this section, the 2×2 global stiffness matrix (which is identical to the element stiffness matrix) is not invertible because the determinant is zero:

$$\det\left(\frac{EA}{L}\begin{bmatrix} +1 & -1 \\ -1 & +1 \end{bmatrix}\right) = \frac{EA}{L}\left((1)(1) - (-1)(-1)\right) = 0 \qquad (3.29)$$

If we restrain one of the nodes, the active stiffness reduces to a 1×1 matrix which can be easily inverted:

$$\left(\frac{EA}{L}\right)^{-1} = \frac{L}{EA} \qquad (3.30)$$

For more complex structures, any instability will result in a singular active stiffness matrix. Only when a sufficient number of rows and columns are eliminated from the global stiffness matrix does the active stiffness matrix become invertible.

3.5 Reducing the Active Stiffness Matrix

For many structural systems, it is possible to infer that certain unknown displacements are either zero or equal in magnitude to other displacements. Mostly, these observations are a result of geometrical **symmetry**. Good practice dictates that we employ this symmetry to try to reduce our active stiffness matrix in order to lower the computational burden of inverting the active stiffness matrix. We can achieve this result by modelling only a portion of a structure or reducing the active stiffness matrix. Both approaches require that we arrive at an active stiffness matrix defined in size by the number of **unique, nonzero active dofs**. In order to show how to reduce the size of an active stiffness matrix, consider the following example for which we immediately number the active/natural dofs:

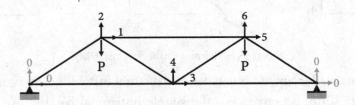

Figure 3.19.   Example structural problem with active dofs numbered.

By inspection, we note that the structural problem (structure, loads, and supports), is vertically symmetrical. This symmetry can be used to establish three relationships between the unknown displacements:

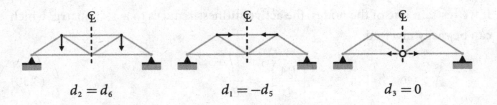

$$d_2 = d_6 \qquad\qquad d_1 = -d_5 \qquad\qquad d_3 = 0$$

Figure 3.20.   Relationships between unknown displacements due to symmetry.

This system demonstrates three basic rules of symmetry: 1) displacements acting parallel to the line of symmetry are equal; 2) displacements acting perpendicular to the line of symmetry are equal and opposite; and 3) displacements lying on the line of symmetry and acting perpendicular to the line of symmetry are zero.

Because all prescribed displacements in our example are zero, the active stiffness equation can be simplified:

$$\{P_U\} = [K_{UU}]\{d_U\} \tag{3.31}$$

We can rewrite this matrix equation explicitly as a set of linear equations:

$$
\begin{aligned}
P_1 &= K_{11}d_1 + K_{12}d_2 + K_{13}d_3 + K_{14}d_4 + K_{15}d_5 + K_{16}d_6 \\
P_2 &= K_{21}d_1 + K_{22}d_2 + K_{23}d_3 + K_{24}d_4 + K_{25}d_5 + K_{26}d_6 \\
P_3 &= K_{31}d_1 + K_{32}d_2 + K_{33}d_3 + K_{34}d_4 + K_{35}d_5 + K_{36}d_6 \\
P_4 &= K_{41}d_1 + K_{42}d_2 + K_{43}d_3 + K_{44}d_4 + K_{45}d_5 + K_{46}d_6 \\
P_5 &= K_{51}d_1 + K_{52}d_2 + K_{53}d_3 + K_{54}d_4 + K_{55}d_5 + K_{56}d_6 \\
P_6 &= K_{61}d_1 + K_{62}d_2 + K_{63}d_3 + K_{64}d_4 + K_{65}d_5 + K_{66}d_6
\end{aligned}
\tag{3.32}
$$

Then, we introduce the symmetry relationships ($d_2 = d_6$, $d_1 = -d_5$ and $d_3 = 0$):

$$
\begin{aligned}
P_1 &= K_{11}d_1 + K_{12}d_2 + K_{13}(0) + K_{14}d_4 + K_{15}(-d_1) + K_{16}(d_2) \\
P_2 &= K_{21}d_1 + K_{22}d_2 + K_{23}(0) + K_{24}d_4 + K_{25}(-d_1) + K_{26}(d_2) \\
P_3 &= K_{31}d_1 + K_{32}d_2 + K_{33}(0) + K_{34}d_4 + K_{35}(-d_1) + K_{36}(d_2) \\
P_4 &= K_{41}d_1 + K_{42}d_2 + K_{43}(0) + K_{44}d_4 + K_{45}(-d_1) + K_{46}(d_2) \\
P_5 &= K_{51}d_1 + K_{52}d_2 + K_{53}(0) + K_{54}d_4 + K_{55}(-d_1) + K_{56}(d_2) \\
P_6 &= K_{61}d_1 + K_{62}d_2 + K_{63}(0) + K_{64}d_4 + K_{65}(-d_1) + K_{66}(d_2)
\end{aligned}
\tag{3.33}
$$

Gathering terms we arrive at a simplified expression:

$$
\begin{aligned}
P_1 &= (K_{11} - K_{15})d_1 + (K_{12} + K_{16})d_2 + K_{14}d_4 \\
P_2 &= (K_{21} - K_{25})d_1 + (K_{22} + K_{26})d_2 + K_{24}d_4 \\
P_3 &= (K_{31} - K_{35})d_1 + (K_{32} + K_{36})d_2 + K_{34}d_4 \\
P_4 &= (K_{41} - K_{45})d_1 + (K_{42} + K_{46})d_2 + K_{44}d_4 \\
P_5 &= (K_{51} - K_{55})d_1 + (K_{52} + K_{56})d_2 + K_{54}d_4 \\
P_6 &= (K_{61} - K_{65})d_1 + (K_{62} + K_{66})d_2 + K_{64}d_4
\end{aligned}
\tag{3.34}
$$

Since we have six equations and three unknowns, three of the equations will be redundant. One way to remove the redundant equations is to only use the equations that relate to unique dofs (1, 2, and 4):

$$
\begin{aligned}
P_1 &= (K_{11} - K_{15})d_1 + (K_{12} + K_{16})d_2 + K_{14}d_4 \\
P_2 &= (K_{21} - K_{25})d_1 + (K_{22} + K_{26})d_2 + K_{24}d_4 \\
P_4 &= (K_{41} - K_{45})d_1 + (K_{42} + K_{46})d_2 + K_{44}d_4
\end{aligned}
\tag{3.35}
$$

These three equations provide a reduced version of the active stiffness equation:

$$
\begin{Bmatrix} P_1 \\ P_2 \\ P_4 \end{Bmatrix} =
\begin{bmatrix}
K_{11} - K_{15} & K_{12} + K_{16} & K_{14} \\
K_{21} - K_{25} & K_{22} + K_{26} & K_{24} \\
K_{41} - K_{45} & K_{42} + K_{46} & K_{44}
\end{bmatrix}
\begin{Bmatrix} d_1 \\ d_2 \\ d_4 \end{Bmatrix}
\tag{3.36}
$$

This method relies on intuition about which relationships are unique. The more reliable approach is to combine equations (rows) together based on the operations used to collapse the columns:

$$
\begin{Bmatrix} P_1 - P_5 \\ P_2 + P_6 \\ P_4 \end{Bmatrix} =
\begin{bmatrix}
K_{11} - K_{15} - K_{51} + K_{55} & K_{12} + K_{16} - K_{52} - K_{56} & K_{14} - K_{54} \\
K_{21} - K_{25} + K_{61} - K_{65} & K_{22} + K_{26} + K_{62} + K_{66} & K_{24} + K_{64} \\
K_{41} - K_{45} & K_{42} + K_{46} & K_{44}
\end{bmatrix}
\begin{Bmatrix} d_1 \\ d_2 \\ d_4 \end{Bmatrix}
\tag{3.37}
$$

Although reducing the active stiffness matrix is not necessary, it does decrease the computational burden of analysis. Knowing which dofs will be identical or negligible is also a good strategy for debugging inputs and verifying results. The reader can use the following set of four example problems to see if they can determine the minimum size of the active stiffness matrix (answers provided in the caption):

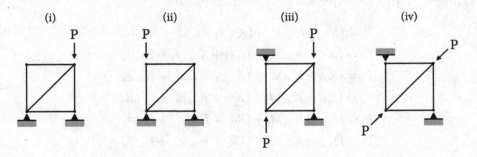

Figure 3.21.   Number of unique, nonzero active dofs: (i) 2x2, (ii) 1x1, (iii) 2x2, (iv) 1x1.

Chapter 4

Matrix Analysis Code

In this chapter, we present computer code for the general matrix analysis method developed in Chapter 3. Although computers are able to perform many simple operations very quickly, they lack the human intuition to discern which calculations are necessary for a given process. Hence, we need to specify the scope and duration of every iterative procedure using numbers and indices. Our general method already accounts for some level of iteration (i.e., iterating through elements to assemble the global stiffness matrix), but we must now ensure that every iterative component of our analysis has a corresponding parameter.

The algorithms developed in this chapter are designed to accommodate the introduction of elements developed further on in this book. Hence, certain parameters may seem superfluous on first inspection, but will become relevant later on in our exploration of MSA and FEM. The code also permits the combination of compatible elements in a single analysis (i.e., a structure composed of both truss and frame elements).

It is important to recognize that we are not developing a user-friendly software package; there are no defaults, warnings, or error messages. Misuse of the code will result in crashed code or nonsensical results, the latter of which is the more dangerous byproduct. The reader must approach the material as both a programmer and a user, treating the code as a tool with a specific range rather than a black box of unlimited capability. This approach also provides the reader with unique insight into the performance of commercial software packages.

While the code structure closely parallels the general method developed in the previous chapter, we need to introduce several additional steps. First, we set up a procedure to define the available set of element functions. Next, we define the set of required inputs and prepare functions to extract additional numbering and

55

indexing required for automation of the analysis. We then present a pair of assembly functions used to generate the active stiffness matrix and force vector. After solving for the unknown displacements, we post-process the results to extract reactions and element forces. These procedures are combined in a general analysis function that can be accessed using an input script. The code is developed with reference to the truss example from the previous chapter.

4.1   Defining the Element Functions

The basic building blocks of matrix analysis are the element functions. With each new element, we must not only code in a new element function, but must also define how these functions are accessed.

4.1.1   Element Functions

The truss function from Chapter 2, [Ke,ke,Te] = **Ke_truss**(xe,prop), is one of seven functions that we develop in this text. Each function name is created by concatenating **Ke_** with the element type (**truss**, **beam**, **frame**, **heat**, **elastic**, **plate**, **shell**) and must maintain the same inputs (xe, prop) and outputs (Ke, ke, Te).

4.1.2   Element Definition Function

In order to be able to introduce new elements without changing the main procedure, we set up the **defElems** function, which must be updated each time we introduce a new element. The code for this function is presented below:

```
 1 function [kList,iad] = defElems
 2
 3 net = 0;                                    % number of element types
 4
 5 % 1. Truss
 6 net = net + 1;
 7 kList{net} = 'Ke_truss';                        % list of function names
 8 iad(net,:,1) = [1 0 0 0 0 0 0];      % index of activated dofs - 1D
 9 iad(net,:,2) = [1 1 0 0 0 0 0];      % index of activated dofs - 2D
10 iad(net,:,3) = [1 1 1 0 0 0 0];      % index of activated dofs - 3D
```

This function takes no inputs, but generates two pieces of information: 1) a list of strings, kList, storing the name of each element function and 2) a three-dimensional matrix, iad, containing the **index of <u>a</u>ctivated <u>d</u>ofs**, identifying the nodal dofs occupied by an element in each of the three spatial dofs. Because the function names will differ in length, kList is initialized as a **cell**, a data structure in MATLAB that can store matrices and arrays of different sizes within the same variable. We will also use cells to store our variably-sized element stiffness matrices. The strings specified in kList must exactly match the names of each element function.

The reader may observe that the index of activated dofs references a total of seven dofs even though the truss element has at most three dofs per node (in three dimensions). We include these seven dofs in anticipation of more advanced elements affected by translations (u, v, w), rotations (θ_x, θ_y, θ_z), and temperature (T). While no single element in this text will occupy all seven dofs (though the frame and shell elements can occupy six out of the seven), there are elements available in advanced literature that combine these dofs in other permutations. Furthermore, these seven degrees of freedoms are common in structural and mechanical applications, thus providing the reader with a robust template for a wide range of structural analyses problems. Defining each element's dof contributions (and their order) will also make it possible to combine different types of elements in more complex, hybrid structures.

To add a new element, we replicate lines 5 through 10 and then modify the new entries for kList and iad according to the specifics of the associated element function. It is worth noting that neither output is initialized; instead, kList and iad are both built up with the addition of each element. This approach is generally considered bad practice, but it does reduce the number of variables to track when adding a new element. While MATLAB is able to initialize and resize variables inside the main code, other languages do not permit this flexibility; the reader should be aware of this limitation in any attempts to appropriate this code to another language.

4.2　Defining the Inputs

In order to set up a problem for code implementation, we must be precise and complete in defining its parameters. For matrix analysis, we need to specify a total of six inputs: 1) nodal coordinates, 2) element connectivity, 3) element properties, 4) index of dofs at BCs, 5) applied forces, and 6) prescribed displacements. We begin by recalling the example from the previous chapter:

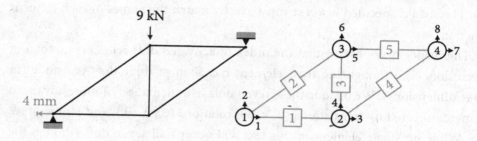

Figure 4.1.　Example problem. For all elements, E = 200,000 MPa.
For elements 1, 3, and 5, A = 100 mm². For elements 2 and 4, A = 200 mm².

Since a computer only processes numbers, it is up to the user to ensure consistent units between inputs and results. For our example, forces are expressed in N while dimensions/displacements are presented in mm. To be consistent, areas are recorded in mm² and moduli/stresses are presented in MPa (N/mm²).

4.2.1　Nodal Coordinates

In order to define the geometry of the structural problem, we must provide both the nodal coordinates and element connectivity. The choice of origin is arbitrary; for our example we use the first node:

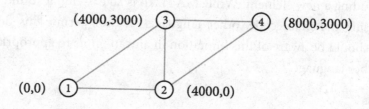

Figure 4.2.　Nodal coordinates in mm.

The nodal coordinates are stored in the **nodal coordinate matrix**, xn:

xn	x	y
1	0	0
2	4000	0
3	4000	3000
4	8000	3000

The number of columns in this matrix defines the number of spatial dimensions, nsd, while the number of rows defines the number of nodal points, nnp; hence, it is very important to dimension this matrix correctly.

At this point, it is worth mentioning that we will present matrices in **row-major order** throughout this text. The **storage order** is a computer science term that defines how multidimensional matrices are stored in linear memory. As suggested by the terminology, row-major order stores matrix values row-by-row, while **column-major order** stores values column-by-column. We demonstrate the two storage orders of our coordinate matrix below:

	row major	column major
0	0	0
1	0	4000
2	4000	4000
3	0	8000
4	4000	0
5	3000	0
6	8000	3000
7	3000	3000

In any given operation, we are more likely to access the set of coordinates for a specific node (i.e., node 2, whose entries are highlighted above) rather than the x-components for all nodes. Hence, our presentation of the nodal coordinate matrix implies that it should be stored in row-major order. However, storage order is code-specific; like many of the languages originally used to develop MSA and FEM, MATLAB is column-major. In this book, we have decided to present matrices in row-major order against convention. This choice is motivated by the assumption that readers are more comfortable with tabulations where properties appear as column headers and iterations index the rows. Though a mismatch in storage order can technically result in slower computation, the difference is negligible unless very large problems are undertaken. This minor inefficiency is deemed a sensible exchange for clarity.

4.2.2 Element Connectivity

Having defined our nodal coordinates, we next need to specify how elements are connected. This information is stored in the **index of element nodes**, ien:

ien	n=1	2
e=1	1	2
2	1	3
3	2	3
4	2	4
5	3	4

The number of rows in the ien matrix defines the number of elements, nel. The number of nonzero entries in each row defines the number of nodes per element, nen. Our code permits us to combine different elements with different numbers of nodes; we must consistently populate each row starting from the left without skipping entries. It is worth noting that MATLAB indexing starts with 1, whereas many other languages start with 0.

4.2.3 Element Properties

While many structural systems are composed of elements with uniform properties, a robust code should accommodate unique elements. We use the **element property matrix** (prop) to store element properties row by row. The first column in this matrix specifies the element type (as defined in the **defElems** function) and subsequent entries define additional properties (as defined inside the individual element functions). To streamline the introduction of elements developed later in the text, the element property matrix has sixteen columns corresponding to the various unique element properties. Because we do not need to pass the element type to individual element functions, the property matrix index is offset by one inside the element function.

		1	2	3	4	5	6	7	8	9	10	11	12	13	14	15	16
	nsd	typ	E	A	I_1	I_2	J	v	t	k	kh	PS	nph	npl	npm	npv	npb
Truss	1-3	1	✓	✓													
Beam	1	2	✓		✓							(✓)						
	2		✓	✓			✓	✓				(✓)						
Frame	1,2	3	✓	✓	✓							(✓)						
	3		✓	✓	✓	✓	✓	✓				(✓)						
Heat	1	4		✓						✓				(✓)				
	2								✓	✓				(✓)				
Elastic	2,3	5	✓					✓	✓				✓		(✓)	(✓)		
Plate	2	6	✓					✓	✓								✓	✓
Shell	2,3	7	✓					✓	✓						✓	✓	✓	✓

For our problem, we only need to provide values in the first three columns:

prop	type	E(MPa)	A(mm^2)	I_1	I_2	J	v	t	k	kh	PS	nph	npl	npm	npv	npb
e=1	1	200000	100	0	0	0	0	0	0	0	0	0	0	0	0	0
2	1	200000	200	0	0	0	0	0	0	0	0	0	0	0	0	0
3	1	200000	100	0	0	0	0	0	0	0	0	0	0	0	0	0
4	1	200000	200	0	0	0	0	0	0	0	0	0	0	0	0	0
5	1	200000	100	0	0	0	0	0	0	0	0	0	0	0	0	0

4.2.4 Index of Degrees of Freedom for BCs

The inputs presented thus far are sufficient to characterize the structure but provide no information about the BCs. To complete the structural problem, we need to specify the type of BC at each global dof, the applied force at each natural dof, and the prescribed displacement at each essential dof. First, we specify the type of BC (natural or essential) at each global dof using the **index of BC dofs** (`idb`):

idb	u	v	w	θ_x	θ_y	θ_z	t
n=1	1	1	0	0	0	0	0
2	0	0	0	0	0	0	0
3	0	0	0	0	0	0	0
4	1	1	0	0	0	0	0

The idb must be dimensioned so that the number of rows exactly matches the number of nodes established in the xn matrix and the number of columns matches the number of dofs established by the iad matrix. The idb matrix is **binary**; a dof marked by 1 specifies an essential BC. All values other than 0 are understood to be 1 and any entries in fields not occupied by a global dof are ignored.

While the more efficient storage method is to use an array the size of the global stiffness matrix, the more user-friendly (and debug-friendly) approach is to set up a matrix with all dofs (as defined by the iad matrix) enumerated for all nodes.

4.2.5 Applied Force Vector

We next define the natural BCs using the **applied force vector**, Pu:

Pu	P_x	P_y	P_z	M_x	M_y	M_z	F_t
n=1	0	0	0	0	0	0	0
2	0	-9000	0	0	0	0	0
3	0	0	0	0	0	0	0
4	0	0	0	0	0	0	0

Although Pu stores the values of a vector, it is dimensioned identically to the idb matrix. Forces applied to restrained/essential dofs and nonexistent dofs (those included in the matrix but not part of the global dofs) are ignored.

4.2.6 Prescribed Displacement Vector

We define the essential BCs using the **prescribed displacement vector**, ds:

ds	u	v	w	θ_x	θ_y	θ_z	t
n=1	-4	0	0	0	0	0	0
2	0	0	0	0	0	0	0
3	0	0	0	0	0	0	0
4	0	0	0	0	0	0	0

The ds matrix must be dimensioned identically to the idb matrix. Prescribed displacements applied at active/natural and nonexistent dofs are ignored.

4.3   Generate Numbers and Indices

Good coding practice dictates that inputs specify the minimum amount of information needed for analysis. Although this strategy reduces the likelihood of inconsistent instructions, it does require that we perform additional operations to extract additional information required for analysis. We will need to generate both **numbers** (defining the number of iterations in a loop), and **indices** (specifying which entries are accessed during a loop).

4.3.1   Generate Numbers

There are a total of eight numbers that we need to generate: net, ndf, nsd, nnp, nel, nen, ned, and neq. The **number of element types**, net, has already been defined in the defElems function; it is updated with the addition of each element but does not otherwise need to be used in any part of the code. The **number of degrees of freedom**, ndf, is implicitly defined in the defElems function, and can always be extracted from the number of rows in iad, idb, Pu, or ds using nsd = size(idb,2). For our example problem, these two numbers are:

net $\boxed{1}$ ndf $\boxed{7}$

We can extract three of these numbers directly from the inputs. The **number of spatial dimensions**, nsd, is defined by the number of columns in the xn matrix, nsd = size(xn,2). The **number of nodal points**, nnp, is defined by the number of rows in the xn, idb, Pu, or ds matrices, nnp = size(xn,1). The **number of elements**, nel, is defined by the number of rows in the ien or prop matrices, nel = size(ien,1). For our example, these three numbers are:

nsd $\boxed{2}$ nnp $\boxed{4}$ nel $\boxed{5}$

There are two element-specific number arrays that we must also extract: number of element nodes and number of element dofs. The **number of element nodes**, nen, can be extracted directly from the connectivity matrix, nen = sum(ien > 0,2). The **number of element dofs**, ned, can be found by counting the number of dofs specified by the iad for a specific nsd and element type. For our example, both arrays will contain the number 2 (2 nodes per element and 2 total dofs per element node):

nen	node
e=1	2
2	2
3	2
4	2
5	2

ned	dof
e=1	2
2	2
3	2
4	2
5	2

Finally, we need to determine the total number of active dofs, also known as the **number of equations**, neq. The number of equations is difficult to extract directly from the inputs; it is easier to find once we have generated some of our index matrices. Recalling from Chapter 3 that our active stiffness matrix was 4×4, we know that the number of equations will be 4:

$$\text{neq} \quad \boxed{4}$$

4.3.2 Generate Indices

In our code, we use a total of seven index matrices: iad, idb, ien, ied, idt, ids, and idu. The first of these, iad, is defined in our **defElems** function, while the next two, idb and ien, are provided as inputs. The other four index matrices, ied, idt, ids, and idu, must be generated internally.

The **index of element dofs**, ied, is an enumerated, element-by-element restructuring of the iad matrix. In order to generate this matrix, we need to first prepare a binary version, which we obtain by providing the array, prop(:,1), as the index of the iad using the MATLAB short form, iad(prop(:,1),:,nsd):

	P_x	P_y	P_z	M_x	M_y	M_z	F_t
e=1	1	1	0	0	0	0	0
2	1	1	0	0	0	0	0
3	1	1	0	0	0	0	0
4	1	1	0	0	0	0	0
5	1	1	0	0	0	0	0

In order to enumerate this binary matrix, we introduce a general count index function, **cntIndex**, which we will use later on in the text:

```
1 function [idnew] = cntIndex(id,byRow)
2
3 if not(exist('byRow','var'))          % if byRow not provided, set to 0
4   byRow = 0;
5 end
6
7 cnt = 0;                              % define index counter
8 idnew = 0*id;                         % initialize idnew
9 for i = 1:size(id,1)
10  if byRow, cnt = 0; end              % restart counting for each row
11  for j = 1:size(id,2)
12    if id(i,j) > 0
13      cnt = cnt + 1;
14      idnew(i,j) = cnt;
15    end
16  end
17 end
```

The **cntIndex** function is simple to use. If the byRow flag is set to zero (or not provided), all nonzero entries in the index matrix are ordered by column and then by row. This version of the function is used to generate the ids and idu matrices:

id

1	1	1	0
0	1	0	0
1	1	0	0

cntIndex(id,0) →

idnew

1	2	3	0
0	4	0	0
5	6	0	0

If the byRow flag is set to anything other than zero, all nonzero entries in each row are enumerated by column but reset within the row. This version of the function is used to generate the ied matrix:

id

1	1	1	0
0	1	0	0
1	1	0	0

cntIndex(id,1) →

idnew

1	2	3	0
0	1	0	0
1	2	0	0

We can use this function to generate the enumerated ied matrix. Once again, we take advantage of MATLAB short-form, supplying the binary form of the ied matrix to the **cntIndex** function, **cntIndex**(iad(prop(:,1),:,nsd),1),:

ied	u	v	w	θ_x	θ_y	θ_z	t
e=1	1	2	0	0	0	0	0
2	1	2	0	0	0	0	0
2	1	2	0	0	0	0	0
4	1	2	0	0	0	0	0
5	1	2	0'	0	0	0	0

We can extract the number of element dofs directly, ned = max(ied,[],2).

Next, we need to generate the **index of total dofs**, idt, which specifies if any type of BC (essential or natural) exists at a dof. In order to generate this matrix, we need to iterate through each element, check the nodes to which each element is connected, and then ensure that the idt for that node includes the dofs of that element. To perform this computation, we introduce the **addIndex** function:

```
1 function [idt] = addIndex(ien,ied,nen,nnp)
2
3 nel = size(ien,1);                          % number of elements
4 ndf = size(ied,2);              % number of degrees of freedom
5 idt = zeros(nnp,ndf);                    % index of total dofs
6
7 for e = 1:nel
8    for i = 1:nen(e)
9       n = ien(e,i);
10      idt(n,:) = or(idt(n,:),ied(e,:));
11   end
12 end
```

For our example, the idt is trivial:

idt	u	v	w	θ_x	θ_y	θ_z	t
n=1	1	1	0	0	0	0	0
2	1	1	0	0	0	0	0
3	1	1	0	0	0	0	0
4	1	1	0	0	0	0	0

Next, we need to obtain enumerated indices for each of the two types of BCs. To generate the **index of dofs at supports**, ids, we can enumerate the idb. First, we want to remove any entries in the idb that are inconsistent with the idt using the Boolean operator idb = and(idt,idb). Then, it is a simple matter of finding the ids matrix using ids = **cntIndex(idb)**:

ids	u	v	w	θ_x	θ_y	θ_z	t
n=1	1	2	0	0	0	0	0
2	0	0	0	0	0	0	0
3	0	0	0	0	0	0	0
4	3	4	0	0	0	0	0

The **index of dofs at unknown displacements**, idu, is found by counting all of the indices included in the idt excluding the natural BCs specified in idb, idu = **cntIndex(idt - idb)**:

idu	u	v	w	θ_x	θ_y	θ_z	t	dof
n=1	0	0	0	0	0	0	0	
2	1	2	0	0	0	0	0	
3	3	4	0	0	0	0	0	
4	0	0	0	0	0	0	0	

4.4 Generate Element Stiffness Matrices

Our next task is to generate three element-specific matrices: the **global element stiffness matrix**, Ke, the **local element stiffness matrix**, ke, and the **transformation matrix**, Te. Since our code is designed to permit the combination of different elements, we store these matrices in **cells**, a data structure in MATLAB which allows one variable to store variably-sized matrices. We initialize all three cells simultaneously using the deal function, which makes multiple assignments with one command:

```
[Ke,ke,Te] = deal(cell(nel,1));
```

To populate these matrices, we iterate through all of the elements, using e as the counter. For each element, we begin by defining the **element nodal coordinates**, `xe = xn(ien(e,1:nen(e)),:)`. Next, we use the `kList` cell of strings that we defined in the **defElems** function to generate a **function handle**, `kElem = str2func(kList{prop(e,1)})`. This function handle allows us to indirectly reference the element function for each type of element. We generate the element stiffness matrices by using the function handle, `kElem`, supplied with the element nodal coordinates matrix, `xe`, and the property matrix (void of the element type), `prop(e,2:end)`. The full set of iterations is summarized below:

```
for e = 1:nel
  xe = xn(ien(e,1:nen(e)),:);             % extract element coordinates
  kElem = str2func(kList{prop(e,1)});     % function handle for elem
  [Ke{e},ke{e},Te{e}] = kElem(xe,prop(e,2:end));
end
```

For our example, the three unique element stiffness matrices are presented below:

Ke{1&5}

	q=1	2	3	4
p=1	5000	0	-5000	0
2	0	0	0	0
3	-5000	0	5000	0
4	0	0	0	0

Ke{2&4}

5120	3840	-5120	-3840
3840	2880	-3840	-2880
-5120	-3840	5120	3840
-3840	-2880	3840	2880

Ke{3}

0	0	0	0
0	6667	0	-6667
0	0	0	0
0	-6667	0	6667

4.5   Assemble Active Stiffness Matrix and Force Vector

In order to solve for our displacements, we need to assemble the active stiffness matrix and active force vector. These operations require extensive iteration and index matrices, motivating the introduction of several new functions.

4.5.1   Assemble Active Stiffness Matrix

To generate the active stiffness matrix, we need to use the element global stiffness matrices, `Ke`, to populate the active stiffness matrix based on the information stored in the ied, ien, and idu matrices and nen array. Effectively, our function performs both the mapping and the summation from the previous chapter:

$$K_{PQ} = \sum_e K^e_{pq} \tag{4.1}$$

In code form, this equation is presented similarly:

$$K(P,Q) = K(P,Q) + Ke\{e\}(p,q) \tag{4.2}$$

To populate all of the entries in the active stiffness matrix, we need to iterate through the rows (p, P) and columns (q, Q) of each element (e). Since each column/row defines both an element node and a nodal dof, we must use a total of five nested loops (elements, row nodes, row dofs, column nodes, and column dofs). The **addStiff** function assembles the active stiffness matrix, K, from the element stiffness matrices, Ke:

```
 1 function [K] = addStiff(Ke,ien,ied,idu,nen)
 2
 3 nel = size(ien,1);                          % number of elements
 4 ndf = size(ied,2);                          % number of dofs
 5 ned = max(ied,[],2);                 % number of element dimensions
 6 neq = max(idu(:));                        % number of equations
 7 K = zeros(neq);
 8
 9 for e = 1:nel                               % for each element
10   for m = 1:nen(e)                 % for each element node (rows)
11     for i = 1:ndf
12       P = idu(ien(e,m),i);                   % global row index
13       if and(P,ied(e,i))              % check if dof is active
14         for n = 1:nen(e)             % for each element node (cols)
15           for j = 1:ndf
16             Q = idu(ien(e,n),j);               % global col index
17             if and(Q,ied(e,j))          % check if dof is active
18               p = (m-1)*ned(e) + ied(e,i);      % local row index
19               q = (n-1)*ned(e) + ied(e,j);      % local col index
20               K(P,Q) = K(P,Q) + Ke{e}(p,q);
21             end
22           end
23         end
24       end
25     end
26   end
27 end
```

Despite the many nested loops, the code is relatively straightforward. The only tricky parts lie in obtaining the global (P, Q) and local (p, q) indices. The global indices are found using the global nodes (ien(e,m), ien(e,n)) and local dofs (i,j); these operations occur on lines 12 and 16. The local indices are obtained using the local nodes (m,n), the number of element nodes, ned(e), and the local dofs (i,j); these operations are shown on lines 18 and 19. Before adding in any element stiffness contributions, we must ensure that the referenced dof is active and occupied by the element; this check occurs on lines 13 and 17. Also note that the number of equations, neq, is extracted from the idu using neq = max(idu(:)).

To elucidate the algorithm, we demonstrate how the counters and variables are updated for the iterations performed for element 3 of our example. For convenience of presentation, only the first two iterations of the seven available dofs are provided; thus, the ied(e,i) and ied(e,j) checks are omitted.

e	m	i	P	n	j	Q	p	q	ke{e}(p,q)	K(P,Q)
3	1	1	0							
3	1	2	0							
3	2	1	3	1	1	0				
3	2	1	3	1	2	0				
3	2	1	3	2	1	3	3	3	0	5120
3	2	1	3	2	2	4	3	4	0	3840
3	2	2	4	1	1	0				
3	2	2	4	1	2	0				
3	2	2	4	2	1	3	4	3	0	5120
3	2	2	4	2	2	4	4	4	6667	9547

For our example, we generate the 4×4 active stiffness matrix using the function call Kuu = **addStiff**(Ke,ien,ied,idu,nen):

Kuu	1	2	3	4
1	10120	3840	0	0
2	3840	9547	0	-6667
3	0	0	10120	3840
4	0	-6667	3840	9547

4.5.2 Assemble Active Force Vector

Assembling the active force vector parallels the procedure for assembling the active stiffness matrix. There are two main differences: 1) we are assembling a vector, not a matrix, and 2) we must consider contributions both from the prescribed displacements and the applied force vector.

We begin by calculating the element force vectors due to prescribed displacements. The **genForce** function takes five inputs: the element stiffness matrices, Ke; a global displacement vector, d; the index matrices, ien and ied; and the vector storing the number of element nodes, nen. The function cycles through each element to build a set of element displacement vectors, de{e}, which are used to generate the element force vectors in global coordinates, Fe{e}.

```
1 function [Fe,de] = genForce(Ke,d,ien,ied,nen)
2
3 nel = size(ien,1);                              % number of elements
4 ndf = size(ied,2);                              % number of dofs
5 ned = max(ied,[],2);                     % number of element dofs
6 [Fe,de] = deal(cell(nel,1));
7
8 for e = 1:nel                                   % for each element
9   for n = 1:nen(e)                        % for each element node
10    for i = 1:ndf
11      if ied(e,i) > 0
12        p = (n-1)*ned(e) + ied(e,i);              % local index
13        de{e}(p,1) = d(ien(e,n),i);
14      end
15    end
16  end
17  Fe{e} = Ke{e}*de{e};
18 end
```

We can supply various displacement vectors (ds, du, or d), but they must always be structured as nen by dof matrices. Using Fe = **genForce**(Ke,ds,ien,ied,nen), we generate the five force vectors for our example:

Fe	{1}	{2}	{3}	{4}	{5}
p=1	-20000	-20480	0	0	0
2	0	-15360	0	0	0
3	20000	20480	0	0	0
4	0	15360	0	0	0

Once we find the element forces, we need to assemble the contributions into the active force vector. The **addForce** function mirrors the **addStiff** function, but only iterates through the rows of the vector:

```
1 function [F] = addForce(Fe,ien,ied,id,nen)
2
3 nel = size(ien,1);                          % number of elements
4 ndf = size(ied,2);                            % number of dofs
5 ned = max(ied,[],2);              % number of element dimensions
6 F = zeros(max(id(:)),1);
7
8 for e = 1:nel                              % for each element
9   for n = 1:nen(e)                    % for each element node
10     for i = 1:ndf
11       P = id(ien(e,n),i);                      % global index
12       if and(P,ied(e,i))           % check if dof is active
13         p = (n-1)*ned(e) + ied(e,i);            % local index
14         F(P) = F(P) + Fe{e}(p);
15       end
16     end
17   end
18 end
```

We note that the id`u` input in the **addStiff** function now becomes the general id matrix; when we complete our analysis, we can use the **addForce** function to generate all of our element forces just by supplying id`s` instead of the id`u` matrix. For our example, we find the prescribed displacement contributions to the active force vector using, Kus_ds = **addForce**(Fe,ied,ien,idu,nen):

Kus_ds	(N)
1	20000
2	0
3	20480
4	15360

Our final step is to subtract the prescribed displacement contributions from the active force vector. We begin by removing any applied forces that are not applied at active dofs using Pu = (idu > 0).*Pu. Next, we observe that Kus_ds is a neq by 1 array while the active force vector, Pu, is still a nel by ndf matrix. Hence, we introduce the **chgShape** function, which takes any force or displacement, Fd, determines if it is stored as a matrix or array, and transforms the force or displacement into the other format, Fdnew, using the supplied index matrix, id:

```
 1 function [Fdnew] = chgShape(Fd,id)
 2
 3 nnp = size(id,1);                              % number of nodal points
 4 ndf = size(id,2);                                    % number of dofs
 5 expand = all(size(Fd) == size(id));     % check if expansion required
 6
 7 if expand, Fdnew = zeros(max(id(:)),1);
 8 else Fdnew = zeros(nnp,ndf);
 9 end
10
11 for n = 1:nnp
12   for i = 1:ndf
13     if id(n,i) > 0
14       if expand, Fdnew(id(n,i)) = Fd(n,i);
15       else Fdnew(n,i) = Fd(id(n,i));
16       end
17     end
18   end
19 end
```

The **chgShape** function re-indexes both through expansion:

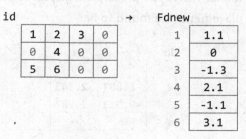

Fd		id				→	Fdnew			
1	1.1	1	2	3	0		1.1	0	-1.3	0
2	0	0	4	0	0		0	2.1	0	0
3	-1.3	5	6	0	0		-1.1	3.1	0	0
4	2.1
5	-1.1
6	3.1

and through reduction:

Fd					id				→	Fdnew	
1.1	0	-1.3	0		1	2	3	0		1	1.1
0	2.1	0	0		0	4	0	0		2	0
-1.1	3.1	0	0		5	6	0	0		3	-1.3
										4	2.1
										5	-1.1
										6	3.1

We use the `chgShape` function to reformat our applied force vector to match our prescribed displacement contributions:

chgShape(Pu,idu)	(N)
1	0
2	-9000
3	0
4	0

The active force vector can now be easily found by taking the difference between the active force vector and the prescribed displacement contributions:

chgShape(Pu,idu) - Kus_ds	(N)
1	0
2	-9000
3	0
4	0

The `chgShape` function is used frequently throughout our code, both in solving for the unknown displacements and post-processing.

4.6   Solve Unknown Displacements

Once we know the active stiffness matrix and the active force vector, determining the unknown displacements is simple:

```
du = chgShape(Kuu\(chgShape(Pu,idu) - Kus_ds),idu);
```

We note that we automatically store the displacement vector in expanded form to match the input prescribed displacements. For our example, the unknown displacements are found to be:

du	u	v	w	θ_x	θ_y	θ_z	t
n=1	0	0	0	0	0	0	0
2	-1.087	-2.343	0	0	0	0	0
3	-0.513	-3.982	0	0	0	0	0
4	0	0	0	0	0	0	0

4.7 Post-Processing

Post-processing is broken down into three parts: 1) calculating the global displacement vector, d; 2) finding the reactions, Rs, and the global force vector, F; and 3) extracting the internal forces.

4.7.1 Global Displacement Vector

Since we have formatted the unknown displacement vector, du, to match the prescribed displacement vector, ds, the global displacement vector, d, is just their sum, d = du + ds, which for our example is found to be:

d	u	v	w	θ_x	θ_y	θ_z	t
n=1	-4.000	0	0	0	0	0	0
2	-1.087	-2.343	0	0	0	0	0
3	-0.513	-3.982	0	0	0	0	0
4	0	0	0	0	0	0	0

4.7.2 Reactions and Global Force Vector

To generate reactions, we first find the element forces using the **genForce** function supplied with the full displacement vector using the function call [Fe,de] = **genForce**(Ke,d,ien,ied,nen). For our example, element forces are found to be:

Fe	{e=1}	{2}	{3}	{4}	{5}
p=1	-14564	-2564	0	-14564	-2564
2	0	-1923	10923	-10923	0
3	14564	2564	0	14564	2564
4	0	1923	-10923	10923	0

For completeness, we also output the element displacements, de:

de	{e=1}	{2}	{3}	{4}	{5}
q=1	-4.000	-4.000	-1.087	-1.087	-0.513
2	0	0	-2.343	-2.343	-3.982
3	-1.087	-0.513	0	0	0
4	-2.343	-3.982	0	0	0

Once we know the element forces, we can find the reactions, Rs, using addForce(Fe,ied,ien,ids,nen). Since the **addForce** function generates an array, we use the **chgShape** function to generate the more comprehensible form:

Rs	P_x	P_y	P_z	M_x	M_y	M_z	F_t
n=1	-17129	-19232	0	0	0	0	0
2	0	0	0	0	0	0	0
3	0	0	0	0	0	0	0
4	17129	10923	0	0	0	0	0

If necessary, we can also find our global force vector, F = Pu + Rs.

4.7.3 Internal Forces

To complete our analysis, we also need to find some element internal forces, Fi, which will depend on the type of element used. Thus, we introduce a **getIntern** function which extracts the relevant internal forces and stresses for each element:

```
1 function [Fi] = getIntern(Fe,de,ke,Te,xn,prop,ien)
2
3 nsd = size(xn,2);                        % number of spatial dimensions
4 nel = size(prop,1);                             % number of elements
5 nen = sum(ien > 0,2);                       % number of element nodes
6 [F,sx] = deal(zeros(nel,1));
7
8 for e = 1:nel
9   typ = prop(e,1);
10  E = prop(e,2);
11  A = prop(e,3);
12  t = prop(e,8);
13
14  xe = xn(ien(e,1:nen(e)),:);               % element nodal coordinates
15  di = Te{e}*de{e};                   % displacements in local coordinates
16  f = Te{e}*Fe{e};                         % forces in local coordinates
17
18  switch typ
19    case 1                                                    % Truss
20      F(e) = -f(1);
21      sx(e) = F(e)/(A*E);
22  end
23 end
24 Fi = cell(2,1);
25 [Fi{:}] = deal(F,sx);
```

With the addition of each new element, this function must be updated to calculate relevant values. These alterations become quite intricate, so we do not document the necessary changes with the addition of each element; the full function can be found in Appendix A. For our analysis, the nonzero entries in the populated internal force vector are found to be:

Fi	F	sx
{e=1}	14564	145.6
{2}	3205	16.0
{3}	-10923	-109.2
{4}	18205	91.0
{5}	2564	25.6

4.8 The Complete Analysis Function

We can group all of our procedures into one general `runAnalysis` function supported by the established functions `defElems`, `cntIndex`, `addIndex`, `addStiff`, `genForce`, `addForce`, `chgShape`, `getIntern`, and the element stiffness functions (i.e., `Ke_beam`). MATLAB requires all of these functions to be stored in the same folder.

The basic structure of the analysis function follows the five steps from the general method presented in the previous chapter: 1) generate numbers and indices, 2) generate element stiffness matrices, 3) assemble active stiffness matrix and active force vector, 4) solve for unknown displacements, and 5) post-process to find the element forces and reactions. Every piece of code up to line 40 has been developed earlier in the chapter; the only new component is the definition of outputs. We use cells to store two types of outputs: `results`, which contain the primary findings of our analysis, and `process`, which contain intermediate matrices and indexes used through the analysis run.

The `runAnalysis` function should not be modified; running a specific analysis or adding new elements is achieved without changing anything in the main code. A new analysis run is initiated by modifying the inputs; a new element is added by updating the `defElems` function, coding in a new element stiffness function, and modifying the `getIntern` function. This analysis function is capable of performing any analysis using elements and techniques covered in this text.

```
1 function [results,process] = runAnalysis(Pu,ds,xn,prop,idb,ien)
2
3 % 0. Define elements
4 [kList,iad] = defElems;
5
6 % 1. Generate numbers and indices
7 nsd = size(xn,2);                          % number of spatial dimensions
8 nnp = size(xn,1);                          % number of nodal points
9 nel = size(ien,1);                         % number of elements
10 nen = sum(ien > 0,2);                      % number of element nodes
11
12 ied = cntIndex(iad(prop(:,1),:,nsd),1);    % index of element dofs
13 idt = addIndex(ien,ied,nen,nnp);           % index of dofs - total
14 idb = and(idt,idb);                        % ensure idb is consistent with idt
15 ids = cntIndex(idb);                       % index of dofs - supports
16 idu = cntIndex(idt - idb);                 % index of dofs - unrestrained
17
18 % 2. Generate element stiffness matrices
19 [Ke,ke,Te] = deal(cell(nel,1));
20 for e = 1:nel
21   xe = xn(ien(e,1:nen(e)),:);              % extract element coordinates
22   kElem = str2func(kList{prop(e,1)});      % function handle for element
23   [Ke{e},ke{e},Te{e}] = kElem(xe,prop(e,2:end));
24 end
25
26 % 3. Assemble active stiffness matrix and active force vector
27 Kuu = addStiff(Ke,ien,ied,idu,nen);        % active stiffness matrix
28 Fe = genForce(Ke,ds,ien,ied,nen);         % element ds contributions
29 Kus_ds = addForce(Fe,ien,ied,idu,nen);    % assembled ds contributions
30
31 % 4. Solve for unknown displacements
32 Pu = (idu > 0).*Pu;          % ensure loads only applied to active dofs
33 du = chgShape(Kuu\(chgShape(Pu,idu) - Kus_ds),idu);
34
35 % 5. Post-process to find element forces and reactions
36 d = du + ds;                               % complete displacement matrix
37 [Fe,de] = genForce(Ke,d,ien,ied,nen);     % element forces and disp
38 Rs = chgShape(addForce(Fe,ien,ied,ids,nen),ids);       % reactions
39 F = Pu + Rs;                               % complete force matrix
40 Fi = getIntern(Fe,de,ke,Te,xn,prop,ien);  % generate internal forces
41
42 % Define outputs
43 results = {F,Rs,Fe,Fi,d,du,de};
44 process = {Kuu,Ke,ke,Te,ied,idu,ids};
```

4.9 Defining the Inputs

There are many ways to set up the inputs for an analysis; we could generate a tabulated input file, use a graphical interface, or simply code in our inputs in a script. For the level of analyses covered in this book, a simple script, such as the example below, is typically the best option:

```
1 % exChpt4 - Analysis of a 2D truss in N and mm base units
2 clear;
3
4 % 1. Global definitions
5 nsd = 2;                            % number of spatial dimensions
6 nel = 5;                                 % number of elements
7 nnp = 4;                              % number of nodal points
8 ndf = 7;                           % number of degrees of freedom
9
10 % 2. Nodal definitions
11 xn = zeros(nnp,nsd);                   % xyz nodal coordinates (mm)
12 xn = [0,0;4,0;4,3;8,3]*1000;
13
14 idb = zeros(nnp,ndf);                  % index of dofs - supported
15 idb(1,:) = 1; idb(4,:) = 1;
16
17 ds = zeros(nnp,ndf);% prescribed displacements (ds) at supports (mm)
18 ds(1,1) = -4;
19
20 Pu = zeros(nnp,ndf);   % applied forces (Pu) at unrestrained dofs (N)
21 Pu(3,2) = -9000;
22
23 % 3. Element definitions
24 ien = zeros(nel,2);                    % index of element nodes
25 ien = [ 1 2; 1 3; 2 3; 2 4; 3 4];
26
27 prop = zeros(nel,3);                        % element properties
28 prop(:,1) = 1;                        % element type (1 = truss)
29 prop(:,2) = 200000;                         % modulus (N/mm^2)
30 prop(:,3) = [100; 200; 100; 200; 100];    % sectional area (mm^2)
31
32 % 4. Run analysis - NOT TO BE CHANGED
33 [results,process] = runAnalysis(Pu,ds,xn,prop,idb,ien);
34 [F,Rs,Fe,Fi,d,du,de] = deal(results{:});
35 [Kuu,Ke,ke,Te,ied,idu,ids] = deal(process{:});
```

We conclude the input script by calling on the **runAnalysis** function and distributing the outputs to their corresponding external variables.

Chapter 5

Beam and Frame Elements

We have used the truss element to demonstrate both how to derive a typical element stiffness equation and how to set up a general structural analysis. In this chapter, we introduce two new elements: the **beam** element, which resolves loads exclusively through bending, and the **frame** element, which combines the mechanical characteristics of the beam and truss elements. In order to derive the beam element stiffness matrix, we first need to establish the constitutive, kinematic, and equilibrium relationships for bending. Once we have formulated the beam stiffness matrix, we introduce the axial dofs to generate a frame element. We then demonstrate how to rotate the beam element into two spatial dimensions and the frame element into both two and three spatial dimensions.

In addition to the general stiffness formulations for the two elements, we develop a **hinge** element, which is able transmit forces but not moments. We provide the code modifications necessary to integrate these new elements into our code and demonstrate their behavior with an example analysis. We conclude the chapter with two supplemental procedures: first, we demonstrate how to account for loads between nodes by using **Equivalent Nodal Loads** (ENLs), and second, we demonstrate how to reduce the size of the active stiffness matrix using symmetry considerations.

5.1 Bending Behavior

Derivation of the element stiffness matrix for the beam follows the same procedure that we used to develop our truss element; we establish constitutive, kinematic, and equilibrium relationships from which we derive the stiffness expression.

5.1.1 Constitutive Relationship

Since we are still concerned with linear elasticity, the constitutive relationship does not change; we refer once again to Hooke's Law:

$$\sigma = E\varepsilon \tag{5.1}$$

5.1.2 Kinematic Relationship

To develop the beam kinematic relationships, we define the 1D beam to lie in the xy-plane, with its neutral axis aligned with the x-axis:

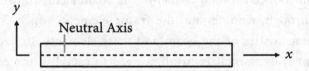

Figure 5.1. Neutral axis.

The **neutral axis** is identified as the position within the cross section where pure bending does not produce strains or stresses.

   Euler-Bernoulli beam theory dictates that the deformation of a beam is fully characterized by the vertical displacement of its neutral axis, $v(x)$. To demonstrate this theory, we examine an infinitesimal portion of the deformed shape:

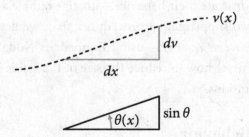

Figure 5.2. Infinitesimal portion of a deformed shape of a beam.

The deformed beam slope, dv/dx, is equal to the sine of the **rotation**, $\theta(x)$:

$$\sin\theta = \frac{dv}{dx} \tag{5.2}$$

Because Euler-Bernoulli beam theory relies on the assumption that deformations and rotations remain small, we can take advantage of the trigonometric principle that the sine of a small angle is approximately equal to the angle itself:

$$\theta(x) \approx \sin\theta = \frac{dv}{dx} \tag{5.3}$$

The derivative of the beam rotation defines the beam **curvature**, $\kappa(x)$:

$$\kappa(x) = \frac{d\theta}{dx} = \frac{d^2v}{dx^2} \tag{5.4}$$

Over an infinitesimal portion of the beam length, dx, the curvature is constant, while the strain varies linearly through the beam depth:

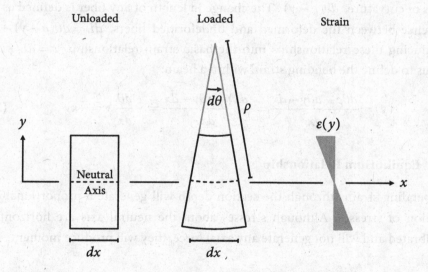

Figure 5.3. Curvature over an infinitesimal portion of the beam.

The deformed neutral axis traces the arc of a circle with **radius of curvature**, ρ. The **change in slope**, $d\theta$, occupies a fraction of a full circle, 2π, which is equal to the ratio occupied by the segment, dx, of the circumference, $2\pi\rho$:

$$\frac{dx}{2\pi\rho} = \frac{d\theta}{2\pi} \tag{5.5}$$

Reorganizing this equation generates the inverse relationship between curvature and the radius of curvature:

$$\kappa = \frac{d\theta}{dx} = \frac{1}{\rho} \tag{5.6}$$

Whereas the strain in a truss is assumed constant through its section, the strain within a beam varies linearly, reaching zero at the neutral axis. Bending strain is thus expressed as a function, $\varepsilon(y)$, of the distance from the neutral axis. In the undeformed configuration, the length of an infinitesimal, longitudinal fiber within the beam will be simply $L = dx$. When the beam experiences positive curvature, the bottom fiber elongates while the top fiber shortens. The length of deformed fibers is directly proportional to the distance from the center of the radius of curvature, $d\theta(\rho - y)$. The change in length of any fiber is defined as the difference between the deformed and undeformed fibers, $dL = d\theta(\rho - y) - dx$. Introducing these relationships into the basic strain relationship, $\varepsilon = dL/L$, allows us to define the bending strain within a beam:

$$\varepsilon(y) = \frac{d\theta\rho - d\theta y - dx}{dx} = \frac{(dx) - d\theta y - dx}{dx} = -\frac{d\theta}{dx} y = -\kappa y \tag{5.7}$$

5.1.3 Equilibrium Relationship

The bending strains through the section depth will generate a proportional distribution of stresses. Although stresses about the neutral axis are horizontally equilibrated and will not generate any axial force, they will produce moment:

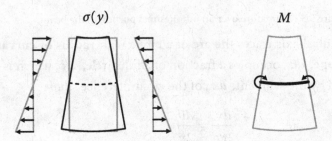

Figure 5.4. Bending stress and moment.

The equilibrating moment is the product of the stress and the distance from the neutral axis integrated over the cross-sectional area. Because positive moment occurs when the top half experiences compression, we use a negative sign:

$$M = -\int \sigma y dA \tag{5.8}$$

By introducing both the kinematic and constitutive relationships into this equation, we obtain the relationship between moment and curvature:

$$M = -\int (E\varepsilon) y dA = \int E(\kappa y) y dA = E\kappa \int y^2 dA \tag{5.9}$$

The integral in this equation defines the **second moment of area**, I, more commonly called the **moment of inertia** in engineering practice:

$$I = \int y^2 dA \tag{5.10}$$

By introducing the moment of inertia into equation (5.9), we obtain a simpler form of the relationship between moment and curvature:

$$M = EI\kappa \tag{5.11}$$

5.2 Stiffness Formulation

In order to use the governing relationships we have developed to generate a beam element stiffness matrix, we need to define element nodes and dofs. While the geometry of a beam element is defined by two nodes, its behavior cannot be fully captured using only two translational dofs. In order to increase the number of dofs so as to adequately model bending behavior, we can either introduce intermediate translational dofs within the element or add rotational dofs at the nodes:

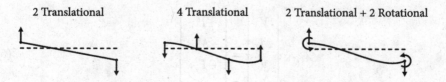

Figure 5.5. Potential beam dofs (only the right two are sufficient to capture bending behavior).

Because four translational dofs make it difficult to combine elements, the traditional approach is to use a translational and rotational dof at each node:

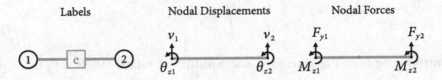

Figure 5.6. Labels for elements, nodes, displacements, and forces.

Just as a truss experiences either positive (tensile) or negative (compressive) forces, internal beam forces follow sign convention; positive shear is generated by a left-hand, upward force, while positive moment make the beam smile. For brevity, we will sometimes refer to nodal forces and moments collectively as *forces*.

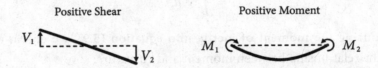

Figure 5.7. Sign conventions for positive internal shears and moments.

Since these forces and moments are the only ones affecting the beam element, they must self-equilibrate. Vertical equilibrium dictates that the two shear forces act in opposite directions, but are equal in magnitude:

$$\sum F_y = 0 = V_1 - V_2; \quad V = V_1 = V_2 \tag{5.12}$$

Moment equilibrium about the first node relates the two end moments:

$$\sum M_z = 0 = -VL - M_1 + M_2; \quad M_2 = M_1 + VL \tag{5.13}$$

The internal forces at the beam ends relate to the global nodal forces as follows:

$$\begin{Bmatrix} V \\ M_1 \\ V \\ M_2 \end{Bmatrix} = \begin{Bmatrix} +F_{y1} \\ -M_{z1} \\ -F_{y2} \\ +M_{z2} \end{Bmatrix} \tag{5.14}$$

5.2.1 Deriving the Beam Stiffness

In order to derive the beam element stiffness, we need to consider both shears and moments. Based on the equilibrium relationships we established in the last section, we know that shear, $V(x)$, will be constant within the beam, while moment, $M(x)$, will vary linearly:

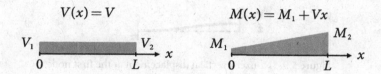

Figure 5.8. Shear and moment distribution within a beam element.

Since curvature is both the derivative of rotation and linearly related to moment, the rotation in the beam is found by integrating the moment equation:

$$\theta_z(x) = \int \kappa\, dx = \int \frac{M}{EI}\, dx = \int \left(\frac{M_1 + Vx}{EI}\right) dx = \frac{V}{2EI}x^2 + \frac{M_1}{EI}x + C_1 \quad (5.15)$$

The rotation at the first node, $\theta_z(0) = \theta_{z1}$, directly supplies the coefficient $C_1 = \theta_{z1}$. The displacements are then found by integrating the rotation:

$$v(x) = \int \theta_z\, dx = \frac{V}{6EI}x^3 + \frac{M_1}{2EI}x^2 + \theta_{z1}x + C_0 \quad (5.16)$$

The deformation at the first node, $v(0) = v_1$, supplies the remaining coefficient $C_0 = v_1$. We summarize the general equations for deformation and rotation:

$$v(x) = \frac{V}{6EI}x^3 + \frac{M_1}{2EI}x^2 + \theta_{z1}x + v_1 \quad (5.17)$$

$$\theta_z(x) = \frac{V}{2EI}x^2 + \frac{M_1}{EI}x + \theta_{z1} \quad (5.18)$$

In order to derive the element stiffness equation, we investigate the system behavior due to the isolated effects of each nodal displacement. Because our derivations are based on linear-elasticity and small-strain assumptions, we will be able to take advantage of the **principle of superposition**, which dictates that the cumulative effect of a set of stimuli on a system is equal to the sum of the effects of the stimuli assessed individually. Thus, we will be able to combine isolated relationships between nodal displacements and forces into one single stiffness expression.

We begin by prescribing a nonzero vertical deformation at the first node, $v_1 \neq 0$, while restraining the other three nodal displacements/rotations, $\theta_{z1} = v_2 = \theta_{z2} = 0$.

Figure 5.9. Nonzero vertical displacement at the first node.

While neither the displacement nor rotation at the first node provides us with new information about the system ($v(0) = v_1$; $\theta_z(0) = \theta_{z1}$), the displacement and rotation at the second node ($x = L$) generates two additional equations:

$$v(L) = 0 = \frac{V}{6EI}(L)^3 + \frac{M_1}{2EI}(L)^2 + v_1 \tag{5.19}$$

$$\theta_z(L) = 0 = \frac{V}{2EI}(L)^2 + \frac{M_1}{EI}(L) \tag{5.20}$$

We solve for the first moment, M_1, and shear, V, in terms of the remaining variables and then find the second moment, M_2, using the moment equation:

$$V = \frac{12EI}{L^3}v_1; \quad M_1 = -\frac{6EI}{L^2}v_1; \quad M_2 = \frac{6EI}{L^2}v_1 \tag{5.21}$$

These three equations provide us with expressions for the four nodal forces/moments as a function of the displacement at the first node:

$$\begin{Bmatrix} F_{y1} \\ M_{z1} \\ F_{y2} \\ M_{z2} \end{Bmatrix} = \begin{Bmatrix} V \\ -M_1 \\ -V \\ M_2 \end{Bmatrix} = \begin{bmatrix} 12EI/L^3 \\ 6EI/L^2 \\ -12EI/L^3 \\ 6EI/L^2 \end{bmatrix} \{v_1\} \tag{5.22}$$

By iterating through each dof, we can determine isolated expressions relating each nodal displacement and rotation to each nodal force and moment:

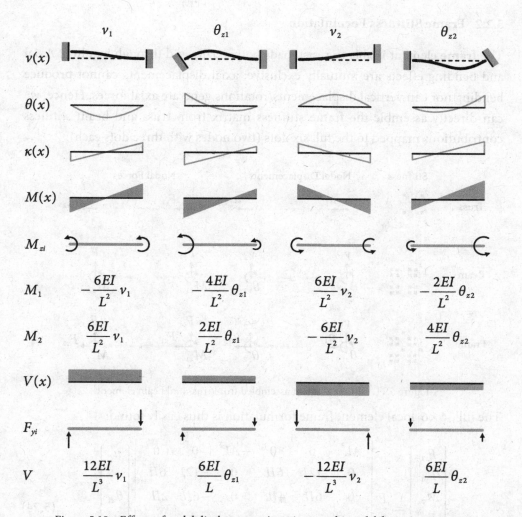

Figure 5.10. Effects of nodal displacements/rotations on the nodal forces/moments.

Using the principle of superposition, we can directly express the dof contributions to each nodal force/moment as the stiffness equation for the beam element:

$$\begin{Bmatrix} F_{y1} \\ M_{z1} \\ F_{y2} \\ M_{z2} \end{Bmatrix} = \frac{EI}{L^3} \begin{bmatrix} 12 & 6L & -12 & 6L \\ 6L & 4L^2 & -6L & 2L^2 \\ -12 & -6L & 12 & -6L \\ 6L & 2L^2 & -6L & 4L^2 \end{bmatrix} \begin{Bmatrix} v_1 \\ \theta_{z1} \\ v_2 \\ \theta_{z2} \end{Bmatrix} \tag{5.23}$$

5.2.2 Frame Stiffness Formulation

The **frame** element is able to resist loads both axially and through bending. Axial and bending effects are mutually exclusive; axial displacements cannot produce bending nor can vertical displacements/rotations generate axial forces. Hence, we can directly assemble the frame stiffness matrix from truss and beam stiffness contributions mapped to the full six dofs (two nodes with three dofs each):

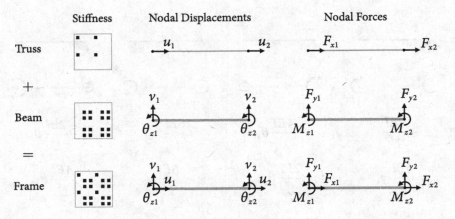

Figure 5.11.   Frame element assembled from truss and beam element.

The full, 6×6 local element frame formulation is thus easily found:

$$\begin{Bmatrix} F_{x1} \\ F_{y1} \\ M_{z1} \\ F_{x2} \\ F_{y2} \\ M_{z2} \end{Bmatrix} = \frac{E}{L^3} \begin{bmatrix} AL^2 & 0 & 0 & -AL^2 & 0 & 0 \\ 0 & 12I & 6IL & 0 & -12I & 6IL \\ 0 & 6IL & 4IL^2 & 0 & -6IL & 2IL^2 \\ -AL^2 & 0 & 0 & AL^2 & 0 & 0 \\ 0 & -12I & -6IL & 0 & 12I & -6IL \\ 0 & 6IL & 2IL^2 & 0 & -6IL & 4IL^2 \end{bmatrix} \begin{Bmatrix} u_1 \\ v_1 \\ \theta_{z1} \\ u_2 \\ v_2 \\ \theta_{z2} \end{Bmatrix} \qquad (5.24)$$

5.2.3 Rotating the Frame Element in 2D

The beam and frame elements derived in the previous sections are local 1D formulations. As with truss elements, we can rotate beam and frame elements into higher dimensions. We start by rotating the frame element into two dimensions:

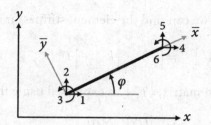

Figure 5.12.   Frame element rotated into two dimensions.

Unlike the truss element, a frame element maintains the same dofs in 1D and 2D:

Local Coordinate System   Global Coordinate System

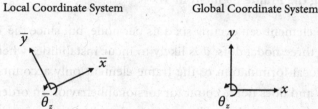

Figure 5.13.   Frame 2D coordinate transformation.

The two translational dofs transform exactly as for our truss element:

$$\begin{aligned} \overline{x} &= +x\cos\varphi + y\sin\varphi \\ \overline{y} &= -x\sin\varphi + y\cos\varphi \end{aligned} \tag{5.25}$$

The rotational dof is unaffected by the rotation from local to global coordinates:

$$\overline{\theta}_z = \theta_z \tag{5.26}$$

We express the coordinate transformation using the frame rotation matrix, $[Q]$:

$$\begin{Bmatrix} \overline{x} \\ \overline{y} \\ \overline{\theta}_z \end{Bmatrix} = [Q] \begin{Bmatrix} x \\ y \\ \theta_z \end{Bmatrix}; \quad [Q] = \begin{bmatrix} +c & +s & 0 \\ -s & +c & 0 \\ 0 & 0 & 1 \end{bmatrix} = \begin{bmatrix} n_{x1} & n_{x2} & 0 \\ n_{y1} & n_{y2} & 0 \\ 0 & 0 & 1 \end{bmatrix} \tag{5.27}$$

We can also use the frame rotation matrix to transform forces and displacements:

$$\begin{Bmatrix} \overline{F}_x \\ \overline{F}_y \\ \overline{M}_z \end{Bmatrix} = [Q] \begin{Bmatrix} F_x \\ F_y \\ M_z \end{Bmatrix}; \quad \begin{Bmatrix} \overline{u} \\ \overline{v} \\ \overline{\theta}_z \end{Bmatrix} = [Q] \begin{Bmatrix} u \\ v \\ \theta_z \end{Bmatrix} \tag{5.28}$$

With these relationships, we can find the element stiffness matrix:

$$\left[K^e\right]_{6\times6} = \left[T^e\right]^T_{6\times6}\left[k^e\right]_{6\times6}\left[T^e\right]_{6\times6} \tag{5.29}$$

The frame transformation matrix, $\left[T^e\right]$, is expressed using the rotation matrix:

$$\left[T^e\right] = \begin{bmatrix} \left[Q\right]_{3\times3} & \left[0\right]_{3\times3} \\ \left[0\right]_{3\times3} & \left[Q\right]_{3\times3} \end{bmatrix} \tag{5.30}$$

5.2.4 Rotating the Frame Element in 3D

In 3D, a frame element can occupy six dofs per node, but since the 1D frame element only has three nodal dofs, it is likely to incur instabilities when it is rotated into 3D. The local formulation of the frame element only accounts for bending about one axis and does not account for torsional behavior. In order to achieve a more stable and more mechanically accurate element, we need to introduce two additional dofs to model out-of-plane bending and one dof to model torsion.

Since we always align the local element axis with the local x-axis, bending can occur in the $\overline{xy}$-plane or the $\overline{xz}$-plane. Our original derivation used the $\overline{xy}$-plane where the z-axis points out of page. In the $\overline{xz}$-plane, the y-axis points into the page, thus reversing the sign of the rotations.

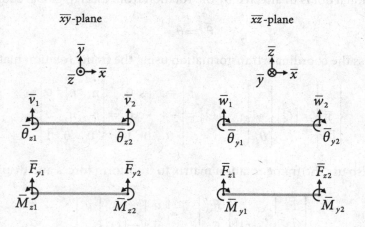

Figure 5.14. Bending in $\overline{xy}$-plane and $\overline{xz}$-plane.

The stiffness equation in the $\overline{xy}$-plane reproduces the original beam equation:

$$\begin{Bmatrix} \overline{F}_{y1} \\ \overline{M}_{z1} \\ \overline{F}_{y2} \\ \overline{M}_{z2} \end{Bmatrix} = \frac{EI_z}{L^3} \begin{bmatrix} 12 & 6L & -12 & 6L \\ 6L & 4L^2 & -6L & 2L^2 \\ -12 & -6L & 12 & -6L \\ 6L & 2L^2 & -6L & 4L^2 \end{bmatrix} \begin{Bmatrix} \overline{v}_1 \\ \overline{\theta}_{z1} \\ \overline{v}_2 \\ \overline{\theta}_{z2} \end{Bmatrix} \tag{5.31}$$

The beam stiffness matrix in the $\overline{xz}$-plane maintains the same values, but reverses the signs of entries whose indices sum to an odd number:

$$\begin{Bmatrix} \overline{F}_{z1} \\ \overline{M}_{y1} \\ \overline{F}_{z2} \\ \overline{M}_{y2} \end{Bmatrix} = \frac{EI_y}{L^3} \begin{bmatrix} 12 & -6L & -12 & -6L \\ -6L & 4L^2 & 6L & 2L^2 \\ -12 & 6L & 12 & 6L \\ -6L & 2L^2 & 6L & 4L^2 \end{bmatrix} \begin{Bmatrix} \overline{w}_1 \\ \overline{\theta}_{y1} \\ \overline{w}_2 \\ \overline{\theta}_{y2} \end{Bmatrix} \tag{5.32}$$

The axial stiffness matches the 1D truss:

$$\begin{Bmatrix} \overline{F}_{x1} \\ \overline{F}_{x2} \end{Bmatrix} = \frac{EA}{L} \begin{bmatrix} +1 & -1 \\ -1 & +1 \end{bmatrix} \begin{Bmatrix} \overline{u}_1 \\ \overline{u}_2 \end{Bmatrix} \tag{5.33}$$

The only remaining dof to include is the rotation about the $\overline{x}$-axis, θ_x, which is resolved through **torsion**. Whereas axial and bending loads typically allow plane sections to remain plane, torsion tends to do the exact opposite by warping the surface. Only sections that are radially symmetric (hollow or full circles) will remain plane when subject to torsion. In this introductory text, we ignore any warping effects due to torsion and maintain our assumption that plane sections remain plane. This simplified view of torsion permits the linear relationship:

$$M_x = \frac{JG}{L} \theta_x \tag{5.34}$$

Here, J is the **torsional constant** and G is the **shear modulus**, which may be expressed as a function of the elastic modulus and **Poisson's ratio**, v:

$$G = \frac{E}{2(1+v)} \tag{5.35}$$

The stiffness equation for torsion closely parallels the axial stiffness equation:

$$\begin{Bmatrix} \overline{M}_{x1} \\ \overline{M}_{x2} \end{Bmatrix} = \frac{JG}{L} \begin{bmatrix} +1 & -1 \\ -1 & +1 \end{bmatrix} \begin{Bmatrix} \overline{\theta}_{x1} \\ \overline{\theta}_{x2} \end{Bmatrix} \tag{5.36}$$

The 3D frame element is found by assembling contributions from axial, $\overline{xy}$-plane bending, $\overline{xy}$-plane bending, and torsion components:

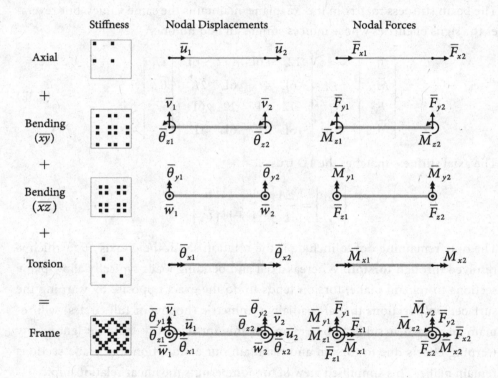

Figure 5.15. Assembly of the frame element from axial, bending, and torsion components.

The unique coefficients for the 3D frame stiffness matrix are supplied using the entries from the truss, bending, and torsional stiffness components:

$$\begin{aligned} k_1 &= AE/L & k_3 &= 6EI_z/L^2 & k_5 &= 12EI_y/L^3 & k_7 &= 4EI_y/L \\ k_2 &= 12EI_z/L^3 & k_4 &= 4EI_z/L & k_6 &= 6EI_y/L^2 & k_8 &= GJ/L \end{aligned} \tag{5.37}$$

These coefficients populate the complete 12×12 frame stiffness matrix as follows:

$$[k^e] = \begin{bmatrix} k_1 & 0 & 0 & 0 & 0 & 0 & -k_1 & 0 & 0 & 0 & 0 & 0 \\ 0 & k_2 & 0 & 0 & 0 & k_3 & 0 & -k_2 & 0 & 0 & 0 & k_3 \\ 0 & 0 & k_5 & 0 & -k_6 & 0 & 0 & 0 & -k_5 & 0 & -k_6 & 0 \\ 0 & 0 & 0 & k_8 & 0 & 0 & 0 & 0 & 0 & -k_8 & 0 & 0 \\ 0 & 0 & -k_6 & 0 & k_7 & 0 & 0 & 0 & k_6 & 0 & k_7/2 & 0 \\ 0 & k_3 & 0 & 0 & 0 & k_4 & 0 & -k_3 & 0 & 0 & 0 & k_4/2 \\ -k_1 & 0 & 0 & 0 & 0 & 0 & k_1 & 0 & 0 & 0 & 0 & 0 \\ 0 & -k_2 & 0 & 0 & 0 & -k_3 & 0 & k_2 & 0 & 0 & 0 & -k_3 \\ 0 & 0 & -k_5 & 0 & k_6 & 0 & 0 & 0 & k_5 & 0 & k_6 & 0 \\ 0 & 0 & 0 & -k_8 & 0 & 0 & 0 & 0 & 0 & k_8 & 0 & 0 \\ 0 & 0 & -k_6 & 0 & k_7/2 & 0 & 0 & 0 & k_6 & 0 & k_7 & 0 \\ 0 & k_3 & 0 & 0 & 0 & k_4/2 & 0 & -k_3 & 0 & 0 & 0 & k_4 \end{bmatrix} \tag{5.38}$$

This local stiffness matrix must be rotated into the global coordinate system. Since the frame element occupies all three translational and all three rotational dofs, we need to perform a full, 3D coordinate transformation. We define the $\bar{x}$-axis of our local ($\overline{xyz}$) coordinate system to align with the element axis:

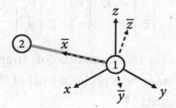

Figure 5.16. Frame rotated into three dimensions.

The element axis defines the first directional cosine, $\{n_{x1}\}$:

$$\{n_x\} = \frac{1}{L}\{x_2 - x_1 \quad y_2 - y_1 \quad z_2 - z_1\} \tag{5.39}$$

Since we do not have enough information for the remaining directional cosines, we have the freedom to define the local $\bar{y}$-axis to lie in the $\overline{xy}$-plane :

$$\{n_y\} = \frac{1}{\sqrt{n_{x1}^2 + n_{x2}^2}}\{-n_{x2} \quad n_{x1} \quad 0\} \tag{5.40}$$

If the element is rotated to align with the z-axis ($n_{x3} = 1$), then the above equation will develop a singularity ($n_{x1} = n_{x2} = 0$). In this case, we set the local $\bar{y}$-axis to align with the global y-axis:

$$\{n_y\} = \{0 \ 1 \ 0\} \tag{5.41}$$

The local $\bar{z}$-axis is found by taking the cross product of the first two axes:

$$\{n_z\} = \{n_x\} \times \{n_y\} \tag{5.42}$$

The 3D frame rotation matrix is assembled out of the three directional cosines:

$$[Q] = \begin{bmatrix} n_{x1} & n_{x2} & n_{x3} \\ n_{y1} & n_{y2} & n_{y3} \\ n_{z1} & n_{z2} & n_{z3} \end{bmatrix} \tag{5.43}$$

This rotation matrix can be used to rotate the coordinate system, translational dofs, and rotational dofs:

$$\begin{Bmatrix} \bar{x} \\ \bar{y} \\ \bar{z} \end{Bmatrix} = [Q] \begin{Bmatrix} x \\ y \\ z \end{Bmatrix} ; \quad \begin{Bmatrix} \bar{u} \\ \bar{v} \\ \bar{z} \end{Bmatrix} = [Q] \begin{Bmatrix} u \\ v \\ z \end{Bmatrix} ; \quad \begin{Bmatrix} \bar{\theta}_x \\ \bar{\theta}_y \\ \bar{\theta}_z \end{Bmatrix} = [Q] \begin{Bmatrix} \theta_x \\ \theta_y \\ \theta_z \end{Bmatrix} \tag{5.44}$$

Since the 3D frame rotation matrix relates two dofs (translational and rotational) at each of the two nodes, it must be repeated a total of four times to populate the 3D frame transformation matrix, $[T^e]$:

$$[T^e] = \begin{bmatrix} [Q]_{3\times3} & [0]_{3\times3} & [0]_{3\times3} & [0]_{3\times3} \\ [0]_{3\times3} & [Q]_{3\times3} & [0]_{3\times3} & [0]_{3\times3} \\ [0]_{3\times3} & [0]_{3\times3} & [Q]_{3\times3} & [0]_{3\times3} \\ [0]_{3\times3} & [0]_{3\times3} & [0]_{3\times3} & [Q]_{3\times3} \end{bmatrix} \tag{5.45}$$

The element stiffness matrix in global coordinates is expressed using the familiar form:

$$\left[K^e\right]_{12\times12} = \left[T^e\right]^T_{12\times12} \left[k^e\right]_{12\times12} \left[T^e\right]_{12\times12} \tag{5.46}$$

5.2.5 Rotating the Beam Element in 2D

The beam element in 2D is a useful complement to the plate element derived later on in the text. This element displaces in the z-axis while maintaining two rotational dofs (about the x- and y-axes):

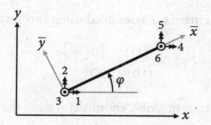

Figure 5.17. Beam rotated into two dimensions.

The local element stiffness matrix is found by adding up contributions from bending in the $\overline{xz}$-plane and torsion about the $\overline{x}$-axis:

$$\begin{Bmatrix} \overline{F}_{z1} \\ \overline{M}_{x1} \\ \overline{M}_{y1} \\ \overline{F}_{z2} \\ \overline{M}_{x2} \\ \overline{M}_{y2} \end{Bmatrix} = \frac{1}{L^3} \begin{bmatrix} 12EI_y & 0 & -6EI_yL & -12EI_y & 0 & -6EI_yL \\ 0 & GJL^2 & 0 & 0 & -GJL^2 & 0 \\ -6EI_yL & 0 & 4EI_yL^2 & 6EI_yL & 0 & 2EI_yL^2 \\ -12EI_y & 0 & 6EI_yL & 12EI_y & 0 & 6EI_yL \\ 0 & -GJL^2 & 0 & 0 & GJL^2 & 0 \\ -6EI_yL & 0 & 2EI_yL^2 & 6EI_yL & 0 & 4EI_yL^2 \end{bmatrix} \begin{Bmatrix} \overline{w}_1 \\ \overline{\theta}_{x1} \\ \overline{\theta}_{y1} \\ \overline{w}_2 \\ \overline{\theta}_{x2} \\ \overline{\theta}_{y2} \end{Bmatrix} \quad (5.47)$$

The transformation from local to global coordinate systems closely mimics the rotation that we used in the 2D frame element, except that there are two in-plane rotations and one out-of-plane displacement:

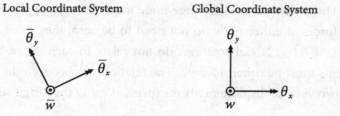

Figure 5.18. Rotation of beam dofs.

The beam rotation matrix is defined as follows:

$$
\begin{Bmatrix} \overline{\theta}_x \\ \overline{\theta}_y \\ \overline{w} \end{Bmatrix} = [Q] \begin{Bmatrix} \theta_x \\ \theta_y \\ w \end{Bmatrix} ; \quad [Q] = \begin{bmatrix} 1 & 0 & 0 \\ 0 & +c & +s \\ 0 & -s & +c \end{bmatrix} = \begin{bmatrix} 1 & 0 & 0 \\ 0 & n_{x1} & n_{x2} \\ 0 & n_{y1} & n_{y2} \end{bmatrix}
\tag{5.48}
$$

The beam transformation matrix is assembled using two rotation matrices:

$$
[T^e] = \begin{bmatrix} [Q]_{3\times3} & [0]_{3\times3} \\ [0]_{3\times3} & [Q]_{3\times3} \end{bmatrix}
\tag{5.49}
$$

The 2D beam stiffness matrix in global coordinates takes the familiar form:

$$
[K^e]_{6\times6} = [T^e]^T_{6\times6} [k^e]_{6\times6} [T^e]_{6\times6}
\tag{5.50}
$$

5.3 Hinge

A hinge is a common structural component that transmits forces, but not moments. We define the hinge as a zero-length condition of the beam or frame element, which we investigate using a simple 1D beam example:

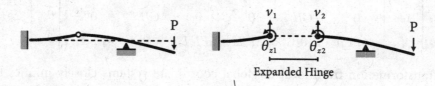

Expanded Hinge

Figure 5.19. Hinge dofs in 1D.

While a hinge occupies one point in space, it is composed of two nodes with two dofs each. The moments at each hinge node must be zero, $M_{z1} = M_{z2} = 0$. Although the forces at either node do not need to be zero, they must be equal and opposite, $F_{y1} = -F_{y2}$. Nodal rotations do not relate to each other, but the nodal displacements must be equal, $v_1 = v_2$. The relationship between the vertical forces and the two vertical displacements is expressed using the **hinge stiffness**, k_h:

$$
F_{y1} = -F_{y2} = -k_h(v_2 - v_1)
\tag{5.51}
$$

Rearranging this equation, we see that the hinge stiffness must be infinitely large:

$$k_h = \frac{F_{y2}}{v_2 - v_1} = \frac{F_{y2}}{(0)} = \infty \tag{5.52}$$

The **hinge stiffness matrix**, $\left[K^{hinge} \right]$, is fully defined using just the hinge stiffness:

$$
\begin{Bmatrix} F_{y1} \\ M_{z1} \\ F_{y2} \\ M_{z2} \end{Bmatrix} = k_h
\begin{bmatrix} 1 & 0 & -1 & 0 \\ 0 & 0 & 0 & 0 \\ -1 & 0 & 1 & 0 \\ 0 & 0 & 0 & 0 \end{bmatrix}
\begin{Bmatrix} v_1 \\ \theta_{z1} \\ v_2 \\ \theta_{z2} \end{Bmatrix} = \left[K^{hinge} \right]
\begin{Bmatrix} v_1 \\ \theta_{z1} \\ v_2 \\ \theta_{z2} \end{Bmatrix} \tag{5.53}
$$

Since we cannot actually supply an infinitely large value for the hinge stiffness, we must use a sufficiently large value to make the difference between the two translational displacements negligible. As we established in the first chapter, structural problems need not be recorded beyond three or four significant figures; thus we need to make sure that the hinge stiffness is more than four orders larger in magnitude than the largest translational stiffness entry in adjacent beam elements. Choosing too large a number, however, can lead to overflow errors. Thus, choosing an appropriate value for the hinge stiffness takes trial and error; it may be better to start with a smaller number and then iterate up by orders of magnitude until the differences between translational displacements are negligible.

Since a hinge can exist in up to three dimensions, we define the general form of the hinge stiffness matrix as follows:

$$
\begin{Bmatrix} F_{x1} \\ F_{y1} \\ F_{z1} \\ F_{x2} \\ F_{y2} \\ F_{z2} \end{Bmatrix} = k_h
\begin{bmatrix} 1 & 0 & 0 & -1 & 0 & 0 \\ 0 & 1 & 0 & 0 & -1 & 0 \\ 0 & 0 & 1 & 0 & 0 & -1 \\ -1 & 0 & 0 & 1 & 0 & 0 \\ 0 & -1 & 0 & 0 & 1 & 0 \\ 0 & 0 & -1 & 0 & 0 & 1 \end{bmatrix}
\begin{Bmatrix} u_1 \\ v_1 \\ w_1 \\ u_2 \\ v_2 \\ w_2 \end{Bmatrix} = \left[K^{hinge} \right]
\begin{Bmatrix} u_1 \\ v_1 \\ w_1 \\ u_2 \\ v_2 \\ w_2 \end{Bmatrix} \tag{5.54}
$$

In each dimension, the hinge stiffness is found by selecting the entries from the 6×6 matrix above. It is important to note that the hinge matrix need not (and should not) be rotated.

5.4 Code Modifications

To integrate these elements into our code, we make some modifications to the defElems function and write new element functions, Ke_beam and Ke_frame.

5.4.1 Modifications to the Element Definition Function

We begin by updating the defElems function. We define two and three active dofs for the 1D and 2D beam elements respectively (we do not need a 3D beam), and three and six dofs for the 1D/2D and 3D frame elements respectively.

```
1 function [kList,iad] = defElems
2
3 net = 0;                                        % number of element types
4
5 % 1. Truss
6 net = net + 1;
7 kList{net} = 'Ke_truss';                        % list of function names
8 iad(net,:,1) = [1 0 0 0 0 0 0];     % index of activated dofs - 1D
9 iad(net,:,2) = [1 1 0 0 0 0 0];     % index of activated dofs - 2D
10 iad(net,:,3) = [1 1 1 0 0 0 0];    % index of activated dofs - 3D
11
12 % 2. Beam
13 net = net + 1;
14 kList{net} = 'Ke_beam';
15 iad(net,:,1) = [0 1 0 0 0 1 0]; % 1D
16 iad(net,:,2) = [0 0 1 1 1 0 0]; % 2D
17 iad(net,:,3) = [0 0 0 0 0 0 0]; % 3D
18
19 % 3. Frame
20 net = net + 1;
21 kList{net} = 'Ke_frame';
22 iad(net,:,1) = [1 1 0 0 0 1 0]; % 1D
23 iad(net,:,2) = [1 1 0 0 0 1 0]; % 2D
24 iad(net,:,3) = [1 1 1 1 1 1 0]; % 3D
```

5.4.2 Beam Element

Next, we introduce the beam element function. We incorporate the hinge element using a conditional statement; if the element length is zero, a hinge is created. Element properties supply E and Iz for the 1D beam, J and v for the 2D beam, and the hinge stiffness, kh, for the hinge.

```
 1 function [Ke,ke,Te] = Ke_beam(xe,prop)
 2
 3 nsd = size(xe,2);                        % number of spatial dimensions
 4 nx = xe(2,:) - xe(1,:);                     % orientation vector, nx
 5 L = norm(nx);                               % beam length, L
 6
 7 if L > 0
 8   nx = nx/L;                                % normalize orientation vector
 9   E = prop(1); I = prop(3);
10   k1 = 12*E*I/L^3; k2 = 6*E*I/L^2; k3 = 4*E*I/L;
11
12   % Initialize local stiffness, ke, and transformation matrix, Te
13   if nsd == 1
14     ke = [k1,   k2,  -k1,   k2;
15            k2,   k3,  -k2,   k3/2;
16           -k1,  -k2,   k1,  -k2;
17            k2, k3/2, -k2,   k3];
18     Qe = [nx,0;0,1];
19   else
20     J = prop(5); v = prop(6); G = E/(2+2*v);
21     k4 = G*J/L;
22     ke = [k1,    0,  -k2, -k1,    0, -k2;
23            0,   k4,    0,   0,  -k4,   0;
24          -k2,    0,   k3,  k2,    0, k3/2;
25          -k1,    0,   k2,  k1,    0,  k2;
26            0,  -k4,    0,   0,   k4,   0;
27          -k2,    0, k3/2,  k2,    0,  k3];
28     Qe = [1,0,0;0,nx(1),nx(2);0,-nx(2),nx(1)];
29   end
30   Te = blkdiag(Qe,Qe);
31 else              % if length is zero, calculate hinge stiffness, kh
32   kh = [eye(nsd,nsd+1); zeros(nsd,nsd+1)];
33   ke = prop(9)*[kh, -kh; -kh, kh];
34   Te = 1;
35 end
36
37 % Find global element stiffness, Ke
38 Ke = Te'*ke*Te;
```

5.4.3 Frame Element

Next, we introduce the frame stiffness function. A similar conditional statement is used to generate the hinge. Element properties provide E, A, and Iz for the 1D/2D frame, Iy, J, and v for the 3D frame, and kh for the hinge.

```
1 function [Ke,ke,Te] = Ke_frame(xe,prop)
2
3 nsd = size(xe,2);                        % number of spatial dimensions
4 nx = xe(2,:) - xe(1,:);                    % orientation vector, nx
5 L = norm(nx);                              % frame length, L
6
7 if L > 0
8   nx = nx/L; E = prop(1); A = prop(2); Iz = prop(3);
9   k1 = E*A/L; k2 = 12*E*Iz/L^3; k3 = 6*E*Iz/L^2; k4 = 4*E*Iz/L;
10
11  % Initialize local stiffness, ke, and transformation matrix, Te
12  if nsd < 3
13    ke = [k1,  0,  0,-k1,  0,  0;
14           0, k2, k3,  0,-k2, k3;
15           0, k3, k4,  0,-k3, k4/2;
16         -k1,  0,  0, k1,  0,  0;
17           0,-k2,-k3,  0, k2,-k3;
18           0, k3, k4/2,0,-k3, k4];
19    if nsd == 1, Qe = [nx 0 0; 0 nx 0; 0 0 1];
20    else         Qe = [nx(1) nx(2) 0; -nx(2) nx(1) 0; 0 0 1]; end
21    Te = blkdiag(Qe,Qe);
22  else
23    Iy = prop(4); J = prop(5); v = prop(6); G = E/(2+2*v);
24    k5 = 12*E*Iy/L^3; k6 = 6*E*Iy/L^2; k7 = 4*E*Iy/L; k8 = G*J/L;
25    ke = [k1,  0,  0,  0,  0,  0,-k1,  0,  0,  0,  0,  0;
26           0, k2,  0,  0,  0, k3,  0,-k2,  0,  0,  0, k3;
27           0,  0, k5,  0,-k6,  0,  0,  0,-k5,  0,-k6,  0;
28           0,  0,  0, k8,  0,  0,  0,  0,  0,-k8,  0,  0;
29           0,  0,-k6,  0, k7,  0,  0,  0, k6,  0, k7/2,0;
30           0, k3,  0,  0,  0, k4,  0,-k3,  0,  0,  0, k4/2;
31         -k1,  0,  0,  0,  0,  0, k1,  0,  0,  0,  0,  0;
32           0,-k2,  0,  0,  0,-k3,  0, k2,  0,  0,  0,-k3;
33           0,  0,-k5,  0, k6,  0,  0,  0, k5,  0, k6,  0;
34           0,  0,  0,-k8,  0,  0,  0,  0,  0, k8,  0,  0;
35           0,  0,-k6,  0, k7/2,0,  0,  0, k6,  0, k7,  0;
36           0, k3,  0,  0,  0, k4/2,0,-k3,  0,  0,  0, k4];
37    if nx(3) == 1, ny = [0 1 0];
38    else          ny = [-nx(2) nx(1) 0]/norm(nx(1:2)); end
39    Qe = [nx;ny;cross(nx,ny)];
40    Te = blkdiag(Qe,Qe,Qe,Qe);
41  end
42 else              % if length is zero, calculate hinge stiffness, kh
43   if nsd < 3, kh = prop(9)*[eye(2,3); zeros(1,3)];
44   else        kh = prop(9)*[eye(3,6); zeros(3,6)]; end
45   ke = [kh, -kh; -kh, kh];
46   Te = 1;
47 end
48 Ke = Te'*ke*Te;                    % Find global element stiffness, Ke
```

5.5 Example

We will use the following braced frame problem to demonstrate the performance of bending elements. Since the structure is composed of frame, truss, and hinge elements, this example also demonstrates the analysis of a hybrid structure.

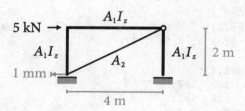

Figure 5.20. Hybrid structure. $E = 200{,}000$ MPa, $I_z = 10^7$ mm^4, $A_1 = 100$ mm^2, $A_2 = 10$ mm^2.

To facilitate this analysis, we need to label elements, nodes, and dofs. We account for the hinge in the top right corner of the frame by defining two nodes, keeping the truss element on the left side of the hinge:

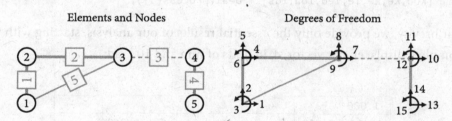

Figure 5.21. Element, node, and dof numbering.

In order to analyze this problem we write a simple input script:

```
1 % exChpt5 - Analysis of a 2D braced frame in N and mm base units
2 clear;
3
4 % 1. Global definitions
5 nsd = 2;                              % number of spatial dimensions
6 nel = 5;                              % number of elements
7 nnp = 5;                              % number of nodal points
8 ndf = 7;                              % number of degrees of freedom
9
10 % 2. Nodal definitions
11 xn = zeros(nnp,nsd);                 % xyz nodal coordinates (mm)
12 xn = [0,0;0,2;4,2;4,2;4,0]*1000;     % xyz nodal coordinates (mm)
13 idb = zeros(nnp,ndf);                % index of dofs - supported
```

```
14 idb(1,:) = 1; idb(5,:) = 1;
15
16 ds = zeros(nnp,ndf);       % prescribed displacements at supports (mm)
17 ds(1,1) = -1;
18
19 Pu = zeros(nnp,ndf);    % applied forces (P) at unrestrained dofs (N)
20 Pu(2,1) = 5000;
21
22 % 3. Element definitions
23 ien = zeros(nel,2);                         % index of element nodes
24 ien = [1 2; 2 3; 3 4; 4 5; 1 3];
25
26 prop = zeros(nel,4);                          % element properties
27 prop = [3 2e5  100  1e7;                      % frame element 1
28         3 2e5  100  1e7;                      % frame element 2
29         3 1e12 0    0 ;                              % hinge
30         3 2e5  100  1e7;                      % frame element 3
31         1 2e5  10   0 ];                         % truss brace
32
33 % 4. RUN ANALYSIS
34 [results,process] = runAnalysis(Pu,ds,xn,prop,idb,ien);
35 [F,Rs,Fe,Fi,d,du,de] = deal(results{:});
36 [Kuu,Ke,ke,Te,ied,idu,ids] = deal(process{:});
```

For brevity, we provide only the essential results of our analysis, starting with the complete displacement vector, d, in units of mm and 10^{-3} rad:

d	u	v	w	θ_x	θ_y	θ_z	t
n=1	-1.000	0	0	0	0	0	0
2	1.491	0.050	0	0	0	-1.368	0
3	1.164	-0.088	0	0	0	0.632	0
4	1.164	-0.088	0	0	0	-0.873	0
5	0	0	0	0	0	0	0

Using these nodal displacements, we can draw the exaggerated deformed shape:

5 kN →

Figure 5.22.   Exaggerated deformed shape.

Next, we summarize the total forces acting on the system in kN and kNm:

F	F_x	F_y	F_z	M_x	M_y	M_z	T
n=1	-4.127	-0.879	0	0	0	4.736	0
2	5.000	0	0	0	0	0	0
3	0	0	0	0	0	0	0
4	0	0	0	0	0	0	0
5	-0.873	0.879	0	0	0	1.746	0

These forces supply us with the ingredients for the FBD:

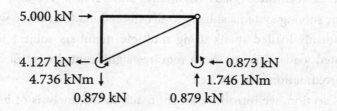

Figure 5.23.   Free Body Diagram.

Finally, we extract the local element forces, once again in units of kN and kNm:

Fe	{e=1}	{2}	{4}	{5}
F	0.500	-1.632	-0.879	0.848
V	3.368	-0.500	-0.873	N/A
M_1	-4.740	2.000	0	N/A
M_2	2.000	0	1.747	N/A

With these element forces, we draw the axial force, shear, and moment diagrams:

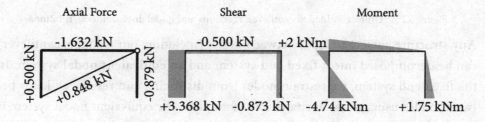

Figure 5.24.   Axial force, shear, and moment diagrams.

5.6   Loads between Nodes

Generally, we want to limit the number of dofs involved in any analysis so as to reduce the computational burden of inverting the active stiffness matrix. Since the number of dofs is controlled by the discretization of the structural system, we are motivated to reduce the number of nodes as much as possible. Based on the techniques we have developed thus far, we must place nodes at all geometrical changes (supports, kinks, element connections, sectional changes, etc.) as well as all applied concentrated loads (moments and forces). We also do not have a method for solving systems subject to distributed loading; while we can approximate uniformly-loaded spans using multiple members subject to approximate concentrated loads, this approach requires significant discretization before adequate approximations are reached.

In this section, we introduce an approach to the analysis of beam and frame systems accounting for loads between nodes. To demonstrate this method, consider a uniformly loaded cantilever, a simple problem for which the analytical reactions and deformations are known:

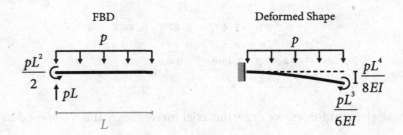

Figure 5.25.   Uniformly loaded cantilever: reactions and nodal displacements/rotations.

Any structure subject to loads between nodes (including our example cantilever) can be decomposed into a **fixed end system** and an **equivalent nodal system**. In the fixed end system, we restrain nodes from displacing and resolve all loads between nodes using classical analytical techniques. The equivalent nodal system is required to equilibrate the forces in the fixed end system. Because the equivalent nodal system experiences only nodal displacements and loads, we are able to analyze it using our established matrix analysis techniques. We can break down our uniformly loaded cantilever into the two contributing systems:

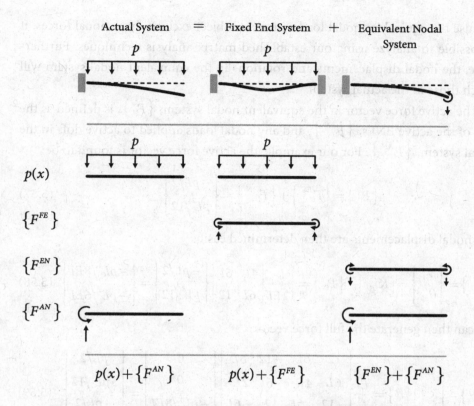

Figure 5.26. Decomposing a structural system into fixed end and equivalent systems.

For our example, the **Fixed End Reactions** (FERs), $\{F^{FE}\}$, are supplied by the solution to the uniformly loaded clamped beam:

$$\{F^{FE}\} = \begin{Bmatrix} pL/2 \\ pL^2/12 \\ pL/2 \\ -pL^2/12 \end{Bmatrix} \tag{5.55}$$

The **Equivalent Nodal Loads** (ENLs), $\{F^{EN}\}$, must equilibrate the FERs:

$$\{F^{EN}\} = -\{F^{FE}\} = \begin{Bmatrix} -pL/2 \\ -pL^2/12 \\ -pL/2 \\ pL^2/12 \end{Bmatrix} \tag{5.56}$$

Because the equivalent nodal load system is subject exclusively to nodal forces, it is possible to analyze using our established matrix analysis techniques. Furthermore, the nodal displacements and rotations in the equivalent nodal system will match those in the actual system.

The active force vector in the equivalent nodal system, $\{P_U\}$, is defined as the sum of the active ENLs, $\{F_U^{EN}\}$, and any nodal loads applied to active dofs in the actual system, $\{F_U^{AN}\}$. For our example, the active force vector is found to be:

$$\{P_U\} = \{F_U^{EN}\} + \{F_U^{AN}\} = \begin{Bmatrix} -pL/2 \\ pL^2/12 \end{Bmatrix} \tag{5.57}$$

The nodal displacements are then determined easily:

$$\{d_U\} = \begin{Bmatrix} v_2 \\ \theta_{z2} \end{Bmatrix} = [K_{UU}]^{-1}\{P_U\} = \frac{L}{12EI}\begin{bmatrix} 4L^2 & 6L \\ 6L & 12 \end{bmatrix}\begin{Bmatrix} -pL/2 \\ pL^2/12 \end{Bmatrix} = \begin{Bmatrix} -pL^4/8EI \\ -pL^3/6EI \end{Bmatrix} \tag{5.58}$$

We can then generate the full force vector:

$$\{F\} = [K^G]\{d\} = \frac{EI}{L^3}\begin{bmatrix} 12 & 6L & -12 & 6L \\ 6L & 4L^2 & -6L & 2L^2 \\ -12 & -6L & 12 & -6L \\ 6L & 2L^2 & -6L & 4L^2 \end{bmatrix}\begin{Bmatrix} 0 \\ 0 \\ -pL^4/8EI \\ -pL^3/6EI \end{Bmatrix} = \begin{Bmatrix} pL/2 \\ 5pL^2/12 \\ -pL/2 \\ pL^2/12 \end{Bmatrix} \tag{5.59}$$

The actual system nodal forces are extracted by subtracting out the ENLs from the total forces acting on the equivalent nodal system:

$$\{F^{AN}\} = \{F\} - \{F^{EN}\} = \begin{Bmatrix} pL/2 \\ 5pL^2/12 \\ -pL/2 \\ pL^2/12 \end{Bmatrix} - \begin{Bmatrix} -pL/2 \\ -pL^2/12 \\ -pL/2 \\ pL^2/12 \end{Bmatrix} = \begin{Bmatrix} pL \\ pL^2/2 \\ 0 \\ 0 \end{Bmatrix} \tag{5.60}$$

As long as we have explicit formulations for FERs, we can apply this method to any structural system subject to loads between nodes. Most FERs that we are likely to encounter in structural analysis can be expressed as a combination of distributed loads, concentrated loads, and concentrated moments:

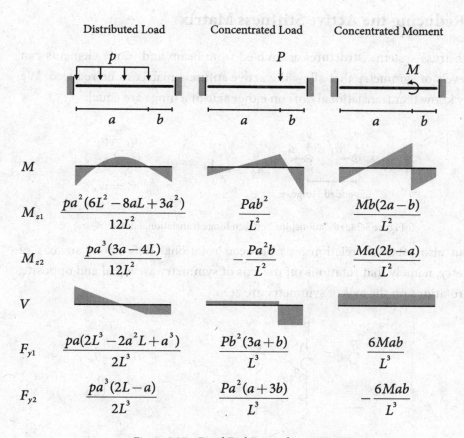

Figure 5.27. Fixed End Forces for common loads.

For each element, the FERs in the local coordinate system are found directly by using the values listed in the above table:

$$-\left\{F^{EN}\right\} = \left\{F^{FE}\right\} = \begin{Bmatrix} F_{y1} \\ M_{z1} \\ F_{y2} \\ M_{z2} \end{Bmatrix} \tag{5.61}$$

It is important to recognize that these procedures are not part of the MSA and FEM framework, but are instead supplementary techniques useful for setting up an analysis that can be performed using matrix analysis techniques.

5.7   Reducing the Active Stiffness Matrix

As with truss systems, structures assembled from beam and frame elements can have levels of symmetry that allow the active stiffness matrix to be reduced. We already know that translational dofs on either side of a hinge are equal:

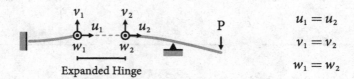

$$u_1 = u_2$$
$$v_1 = v_2$$
$$w_1 = w_2$$

Figure 5.28.   Relationships between hinge translational dofs.

We can also establish relationships between rotational dofs about an axis of symmetry, namely that rotations off the axis of symmetry are equal and opposite, while rotations on the axis of symmetry are zero:

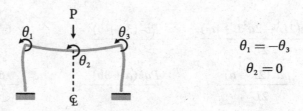

$$\theta_1 = -\theta_3$$
$$\theta_2 = 0$$

Figure 5.29.   Relationships between rotational dofs across a line of symmetry.

Since bending stiffness is typically much lower than axial stiffness, we can frequently ignore axial deformations for bending-dominant systems. This condition can lead to a variety of dof relationships as exemplified by the following diagram:

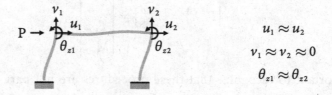

$$u_1 \approx u_2$$
$$v_1 \approx v_2 \approx 0$$
$$\theta_{z1} \approx \theta_{z2}$$

Figure 5.30.   Relationships between dofs when axial deformations are ignored.

Chapter 6

Numerical Approximation

This chapter marks the beginning of our venture into the territory of Finite Element Methods (FEM). While FEM shares the same coding framework as MSA, it follows a different approach to the derivation of elements. In previous chapters, we were able to generate truss, beam, and frame elements using established mechanical principles. Although FEM still relies on classical physical principles, it also requires the use of approximation techniques to transform continuum problems into discretized matrix systems. Since numerical approximation is outside typical academic curricula, we devote this chapter to a general introduction to numerical approximation of boundary value problems.

A **boundary value problem** (BVP) is produced by combining a differential equation with boundary conditions (BCs). The solution to a BVP is a function that not only satisfies the governing differential equation at every point within a domain, but also satisfies the BCs at every point along the domain's boundary. As with our investigation of MSA, there are two types of BCs: essential and natural. In order to keep the problem from being either under or over prescribed, exactly one type of BC must be specified at each point along the boundary.

All of the problems we solved in MSA are more generally classified as BVPs. The essential and natural BCs we used in MSA align exactly with the essential and natural BCs in the BVP. The governing equations that we used to derive our MSA elements such as the truss element,

$$\sigma = E\varepsilon; \quad \varepsilon = \Delta L/L; \quad F = A\sigma \tag{6.1}$$

are actually simplified versions of general 3D governing differential equations:

$$\sigma_{ij} = D_{ijkl}\varepsilon_{kl}; \quad \varepsilon_{kl} = \frac{1}{2}\left(\frac{\partial u_l}{\partial x_k} + \frac{\partial u_k}{\partial x_l}\right); \quad f_i + \frac{\partial \sigma_{ij}}{\partial x_j} = 0 \tag{6.2}$$

111

While we will wait to investigate the continuum mechanics behind these equations, we want to recognize that our derivation of MSA elements relied on simplifications of general relationships: for trusses, all strains and stresses other than the normal strains and stresses along the axis were assumed to be zero; for beams, plane sections were assumed to remain plane. Neither assumption holds true to the general theory, but both simplifications are suitably accurate for many engineering applications.

This type of simplification is among several techniques we use to make our analysis manageable and efficient. We model our elements as prisms of constant section and require that hinges do not transmit any moments; both are *idealizations* of physical reality. Our truss and beam derivations are based on *simplifications* of the governing elasticity differential equations, which are themselves *generalizations* of material and mechanical behavior. All of these techniques are distinct from the type of *approximations* we will be using in solving BVPs. **Numerical approximation** is a technique for solving mathematical problems distinct from the idealizations, simplifications, and generalizations used to transform physical phenomena into their mathematical representations.

Approximation techniques allow us to find numerical, rather than analytical, solutions to mathematical problems. For instance, the trivial problem $x^2 = 2$ has the exact analytical solution $x = \pm\sqrt{2}$ and a numerical approximation $x \approx \pm 1.414$. The benefit of an analytical solution is that it maintains nearly all of the information of an exact solution, allowing us to extract approximate solutions to various degrees of accuracy. The significant limitation of an exact solution is that it can only be obtained using symbolic manipulation. By comparison, we have a myriad of approaches for finding approximate solution; for our simple example, we can use geometric analysis, iterative convergence methods, or even simple trial and error. In fact, as long as we can demonstrate a suitable level of accuracy, any numerical value qualifies as an approximate solution.

BVPs obviously pose a more challenging problem than a simple quadratic equation. In fact, the problems typically analyzed using FEM are so complex that analytical techniques are either too time intensive or practically impossible to obtain. In order to solve such challenging problems, we must develop robust and rigorous approximation approaches.

6.1 Exact Solution to the BVP (Strong Form)

Before we introduce techniques for solving BVPs approximately, we need to establish a baseline for how BVPs are solved exactly. The classical or **strong form** of a BVP produces an exact solution which meets two criteria: 1) the exact solution satisfies the governing differential equation at every point within the **domain**, Ω, and 2) the exact solution satisfies a **boundary condition** (**BC**) at each and every point along the **boundary**, Γ.

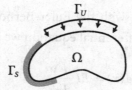

Figure 6.1. Boundary value problem setup.

Every point along the boundary must be assigned either an **essential** (S) or **natural** (U) BC. Essential BCs restrain the function (corresponding to prescribed displacements and temperatures); natural BCs restrain derivatives of the function (corresponding to applied loads, fluxes, and tractions). Exactly one type of BC must be prescribed at every point along the boundary; if two types of BCs are specified at a single point, we prioritize the essential BC. If no BC is specified, we assume that a natural BC of zero magnitude exists. Since BVPs can be constructed in any dimension (1, 2, or 3), it is necessary to adjust the geometry of both the domain (line, area, or volume) and the boundary (point, curve, or surface).

To demonstrate how to find solutions (both exact and approximate) to BVPs, consider the following 2D elasticity problem:

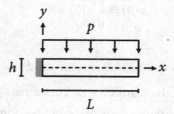

Figure 6.2. Uniformly-loaded cantilever as a 2D problem.

This problem effectively represents a uniformly-loaded cantilever; if we assume that plane sections remain plane, we can simplify the problem as a 1D beam:

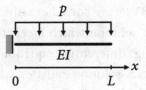

Figure 6.3. Uniformly-loaded cantilever.

This simplification is consistent with the Euler-Bernoulli beam theory from the last chapter, whose governing differential equation we have already established:

$$M(x) = EI \frac{d^2 v}{dx^2} \tag{6.3}$$

To take advantage of this simplification, we must also adapt the BCs:

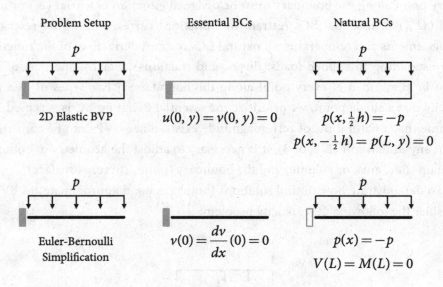

Figure 6.4. Simplified essential and natural BCs.

The fixity at the left side of the cantilever defines a two-part essential BC that restrains both vertical and horizontal displacements along the supported edge of the beam. The first part of this BC is easily achieved by setting the vertical displacement at the support to zero:

$$v(0) = 0 \tag{6.4}$$

The second condition effectively requires that the cross-sectional plane does not rotate. We adapt this condition to the Euler-Bernoulli theory by restraining the fixed-end rotation (approximated by the derivative of the vertical displacement):

$$\frac{dv}{dx}(0) = 0 \tag{6.5}$$

We note that the Euler-Bernoulli bending theory necessitates that essential BCs constrain both the function and its first derivative. An essential BC which restricts the first derivative of the primary function is unusual, but it is a direct consequence of the simplification that planes must remain plane. Restrictions on the rotation of the plane actually restrain the horizontal displacement of the beam plane, which is still a primary function of the general elasticity equations.

The natural BCs in our example are defined by the applied distributed load as well as the null shear and moment at the free end. To simplify our algebra, we will incorporate these three conditions into a single moment equation. We begin by recalling that shear is the first integral of the distributed load:

$$V(x) = \int p(x)\,dx = -p \int dx = -px + c_1 \tag{6.6}$$

Using the free end condition, $V(L) = 0$, we solve for shear explicitly:

$$V(x) = -p(x - L) \tag{6.7}$$

We next obtain the equation for moment by integrating the shear:

$$M(x) = \int V(x)\,dx = -p \int (x - L)\,dx = -p\left(\frac{x^2}{2} - Lx\right) + c_0 \tag{6.8}$$

Using the free end condition again, $M(L) = 0$, we find the moment explicitly:

$$M(x) = -\frac{p}{2}(L - x)^2 \tag{6.9}$$

Combining equations (6.3) and (6.9) provides us with the simplified governing differential equation for our problem:

$$EI\frac{d^2v}{dx^2} = -\frac{p}{2}(L-x)^2 \tag{6.10}$$

For our example, the exact solution, $v(x)$, is now easy to find; all we need to do is reorder the above differential equation and perform two integrations:

$$v(x) = -\frac{p}{2EI}\int\left(\int(L-x)^2 dx\right)dx \tag{6.11}$$

Upon incorporating the essential BCs, we find that the exact solution is quartic:

$$v(x) = -\frac{p}{24EI}\left(x^4 - 4Lx^3 + 6L^2x^2\right) \tag{6.12}$$

We note that this solution exactly satisfies the governing differential equation and both BCs (essential and natural). We will use this exact solution as a benchmark for our approximate solutions.

6.2 Approximate Solution to the BVP (Weak Form)

Since most engineering BVPs are significantly more complicated than our example beam problem, exact solutions are frequently too difficult or even impossible to find using analytical techniques. The examples in this book contradict this statement; in fact, we specifically choose problems for which exact solutions are known to demonstrate the capacity of our approximation techniques. Generally, a good approximate solution is a practical alternative to an unattainable exact solution. Unless an approximate solution replicates the exact solution, it will not satisfy the strong form of the BVP and thus we cannot rely on the classical technique for solving a BVP. Instead, we must introduce an alternate, but still rigorous, approach to the approximate solution of BVPs. This approach requires that we introduce two new concepts: a trial function and a weak form of the BVP.

While there is typically only one exact solution to a BVP, there exist an infinite number of approximate solutions. In fact, any function can be considered an approximate solution to a BVP, but it may not be a sufficiently *good* approximation. In our search for good approximate solutions, we can no longer rely on the BVP to lead us to a solution; we must a priori restrict our solutions to a convenient subset. In FEM, the set of potential solutions are limited to those that can be created using a **trial function**, $\tilde{u}(x)$, defined as the sum of the products of predefined **basis functions**, $N_i(x)$, and unknown **degrees of freedoms (dofs)**, d_i:

$$\tilde{u}(x) = \sum_{neq} d_i N_i(x) \qquad (6.13)$$

By structuring our trial function in this manner, we can span continuous regions by only specifying a set of scalars (dofs). We control the behavior of our trial function by defining a convenient set of linearly independent basis functions; it should be impossible to assemble any basis function from a linear combination of the remaining basis functions. This rule ensures that each dof is an independent variable. By choosing simple polynomials for basis functions, we are able to manipulate the trial function more easily in our future derivations.

Though our example beam problem is not a good demonstration of this complexity, it does provide us with access to the exact solution. Thus, we know that a cubic trial solution will be unable to replicate the quartic exact solution:

$$\tilde{u}_y(x) = \tilde{v}(x) = \sum_{i=0:3} d_i x^i = d_3 x^3 + d_2 x^2 + d_1 x + d_0 \qquad (6.14)$$

We note that the four basis functions (x^3, x^2, x, 1) are linearly independent.

Even in the weak form of the BVP, approximate solutions must still satisfy the essential BCs (at least as well as the form of trial solution allows). Introducing the two essential BCs for our problem, we find that two coefficients are zero:

$$d_1 = d_0 = 0 \qquad (6.15)$$

The trial function is thus reduced to two nonzero dofs:

$$\tilde{v}(x) = d_3 x^3 + d_2 x^2 \qquad (6.16)$$

We plug this trial function directly into our differential equation:

$$6EId_3 x + 2EId_2 = -\frac{p}{2}(L-x)^2 \qquad (6.17)$$

We then gather terms associated with the remaining basic functions:

$$-\frac{p}{2}x^2+\left(pL-6EId_3\right)x+\left(\frac{pL^2}{2}+2EId_2\right)=0 \qquad (6.18)$$

We observe that we do not have an obvious strategy for finding our remaining unknown dofs. Ideally, we would like to find a set of dofs such that the above equation is satisfied at all points within the domain, but with only two remaining dofs such a solution is not possible. We could select any two points along the length of the beam to create two linear equations from which we could obtain our unknown equations, but this choice would be arbitrary. We need to establish a method of finding our unknown dofs in a systematic and objective fashion.

The strong form of the BVP is not particularly useful for solving the remaining dofs since it can only assess if a candidate solution is exact; it cannot evaluate how closely an approximate solution comes to satisfying the BVP. If we establish criteria for evaluating the aptitude of trial solutions, we then have justification for choosing the best among potential approximate solutions. In this chapter, we will demonstrate how the Variational Principle and the Method of Weighted Residuals define metrics for evaluating the aptitude of an approximation and how each approach may be used to generate approximate solutions.

6.3 The Variational Principle (VP)

The **Variational Principle** (VP) is based on the concept that a function that minimizes a system's potential approaches the exact solution. Implementation of the VP follows two steps: first, we determine the potential of the system, and second, we find the trial function (i.e., the set of dofs) that minimizes this potential.

6.3.1 Finding the Total Potential Energy

The VP for elasticity is based on the **principle of least action**, which states that equilibrium is achieved when the total potential energy of a system is minimized. The **total potential energy**, Π, of an elastic system is defined as the difference between the **internal elastic energy**, U, and the **work performed** on the system, W, by both internal body forces, W_Ω, and applied external surface tractions, W_Γ:

$$\Pi = U - W = U - \left(W_\Omega + W_\Gamma \right) \tag{6.19}$$

The total internal elastic energy is found by integrating the **strain energy density**, dU/dV, over the volume of our domain. Strain energy density is defined as the integral of stress with respect to strain:

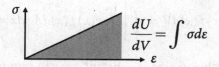

Figure 6.5. Strain energy density under the stress-strain curve.

Since the stress-strain curve in linear elasticity is a straight line, the strain energy density is simply found as the area of the triangle under the stress-strain curve:

$$\frac{dU}{dV} = \int \sigma d\varepsilon = \int E\varepsilon d\varepsilon = \tfrac{1}{2} E\varepsilon^2 \tag{6.20}$$

For bending in a beam, the total internal energy can be decomposed into an integral over the beam cross-sectional area and the beam length:

$$U = \int_\Omega \frac{dU}{dV} dV = \int_L \left(\int_A \frac{dU}{dV} dA \right) dx \tag{6.21}$$

The area integral can be expressed as a function of the local curvature:

$$\int_A \frac{dU}{dV} dA = \int_A \left(\tfrac{1}{2} E\varepsilon^2 \right) dA = \int_A \left(\tfrac{1}{2} E \left(y\kappa \right)^2 \right) dA = \tfrac{1}{2} E\kappa^2 \int_A y^2 dA = \tfrac{1}{2} EI\kappa^2 \tag{6.22}$$

Recalling that curvature is the second derivative of vertical displacement, we find an explicit formulation of the beam strain energy:

$$U = \int_L \left(\int_A \frac{dU}{dV} dA \right) dx = \int_L \left(\tfrac{1}{2} EI\kappa^2 \right) dx = \tfrac{1}{2} EI \int_L \left(\frac{d^2 v}{dx^2} \right)^2 dx \tag{6.23}$$

Next, we look at **work**, generally defined as the product of a force applied at a point and the magnitude of any displacement of that point in the direction of the applied force. We decompose the total work performed on the system into contributions from internal and external forces. The internal work performed on a system is generated by **internal body forces**, f, such as gravity. The internal work performed on our beam is found by taking the volumetric integral of the product of the vertical displacement, $v(x)$, and a vertical body force, $f(x)$:

$$W = \int_{\Omega} v(x)f(x)dV = \int_{L}\left(\int_{A} v(x)f(x)dA\right)dx = A\int_{L} v(x)f(x)dx \qquad (6.24)$$

If we assume that beam area is constant, we can remove it from the integral. Since self-weight is ignored in our example, there is no internal work contribution.

Finally, we look at the external work performed on the system by applied forces. A beam can experience moments as well as concentrated and distributed forces, all of which must be included in the external work expression:

$$W = \sum_{\forall M} M\frac{dv}{dx}(x_M) + \sum_{\forall P} Pv(x_P) + \int_{L} v(x)p(x)dx \qquad (6.25)$$

Work generated by forces is defined as the product of the force and the vertical displacement at the point of the applied force; work generated by moments is defined as the product of the moment and the rotation at the point of the applied moment. Concentrated loads are summed while distributed loads are integrated. For our example, the external work must account for the applied distributed load. Combining the internal and external forms of energy, we obtain the total potential energy expression for our problem:

$$\Pi = \tfrac{1}{2} EI \int_{L}\left(\frac{d^2\tilde{v}}{dx^2}\right)^2 dx - \int_{L} \tilde{v}(x)p(x)dx \qquad (6.26)$$

We next introduce the trial function and its second derivative:

$$\Pi = \tfrac{1}{2} EI \int_{L}\left(6d_3x + 2d_2\right)^2 dx + p\int_{L}\left(d_3x^3 + d_2x^2\right)dx \qquad (6.27)$$

Upon executing the definite integral, we obtain the explicit polynomial expression for the potential of our example problem:

$$\Pi = 6EIL^3d_3^2 + 6EIL^2d_3d_2 + 2EILd_2^2 + pL^4d_3/4 + pL^3d_2/3 \qquad (6.28)$$

6.3.2 Minimizing the Potential

In an elastic system, the total potential energy approaches a minimum as the system arrives at equilibrium. Even though an approximate solution may not be able to reach the same minimum as an exact solution, we can approach the exact solution by minimizing the total potential with respect to each unknown dof:

$$\frac{\partial \Pi}{\partial d_j} = 0 \qquad (6.29)$$

A byproduct of this approach is that we generate as many equations as unknown dofs. For our example, we take the partial derivative with respect to the two remaining unknown dofs (d_2 and d_3) to obtain two equations:

$$\frac{\partial \Pi}{\partial d_2} = 0 = 6EIL^2d_3 + 4EILd_2 + \frac{pL^3}{3}$$

$$\frac{\partial \Pi}{\partial d_3} = 0 = 12EIL^3d_3 + 6EIL^2d_2 + \frac{pL^4}{4} \qquad (6.30)$$

Following a bit of algebraic manipulation, we find the two coefficients:

$$d_3 = \frac{pL}{12EI}; \quad d_2 = -\frac{5pL^2}{24EI} \qquad (6.31)$$

For more complex problems, we will be working with many more dofs which will produce significantly larger sets of equations. If we are strategic in our choice of basis functions, the resulting equations will be linear. The matrix techniques we explored in MSA will prove to be particularly well-suited for solving the resulting linear sets of equations.

6.4   Method of Weighted Residuals (MWR)

The Method of Weighted Residuals (MWR) offers an alternate strategy for finding unknown dofs. Like the VP, the MWR generates a system of linear equations. Unlike the VP, which is based on a physical analogy, the MWR is driven by the mathematical principle of error minimization.

For the beam equation, we begin with a seemingly trivial operation of bringing to one side all of the terms in our governing differential equation:

$$0 = EI\frac{d^2v}{dx^2} - M(x) \tag{6.32}$$

Only an exact solution will satisfy this equation; an approximate solution will introduce an inaccuracy characterized by a **residual**, $R(x)$:

$$R(x) = EI\frac{d^2\tilde{v}}{dx^2} - M(x) \tag{6.33}$$

For our example, we can calculate the residual explicitly:

$$R(x) = 6EId_3x + 2EId_2 + \frac{p}{2}\left(L-x\right)^2 \tag{6.34}$$

The MWR dictates that the exact solution is approached by minimizing the absolute value of the residual. Since the residual is a function, its minimization poses a challenge. One approach is to set to zero the direct integral over the beam length:

$$0 = \int_L R(x)dx \tag{6.35}$$

Unfortunately, this lone equation offers little benefit to solving an arbitrary large set of dofs. In order to generate as many equations as unknown dofs, we multiply the residual by a series of linearly-independent **weighting** functions, $\omega_i(x)$:

$$0 = \int_L \omega_i(x)R(x)dx \tag{6.36}$$

This adjustment affords us additional control over the minimization; we are able to formulate a weighted average that best satisfies the demands of our analysis.

Because the performance of the MWR is significantly dependent on the chosen weighting functions, substantial research has been devoted to their development. In this section, we present the Galerkin and collocation methods, both of which are well-established and reliable techniques.

6.4.1 Galerkin Method

The **Galerkin** method uses the partial derivatives of the trial function with respect to the remaining unknown dofs (which directly produce shape functions):

$$\omega_i(x) = \frac{\partial \tilde{u}}{\partial d_i}(x) = N_i(x) \tag{6.37}$$

For our example, we obtain two equations:

$$0 = \int_L x^2 R(x)dx = \frac{3}{2}EIL^4 d_3 + \frac{2}{3}EIL^3 d_2 + \frac{1}{60}pL^5$$

$$0 = \int_L x^3 R(x)dx = \frac{6}{5}EIL^5 d_3 + \frac{1}{2}EIL^4 d_2 + \frac{1}{120}pL^6 \tag{6.38}$$

Solving this set of equations leads us directly to solutions for the unknown dofs:

$$d_3 = \frac{pL}{18EI}; \quad d_2 = -\frac{3pL^2}{20EI} \tag{6.39}$$

6.4.2 Collocation Method

The **collocation method** employs a set of weight functions defined by the Dirac delta function, δ, evaluated at certain points, x_i:

$$\omega_i(x) = \delta(x - x_i) = \begin{cases} +\infty & x = x_i \\ 0 & x \neq x_i \end{cases} \tag{6.40}$$

Essentially, this method allows us to evaluate the residual at specified points:

$$\int_L \omega_i(x)R(x)dx = \int_L \delta(x - x_i)R(x)dx = R(x_i) \tag{6.41}$$

There are multiple approaches to selecting collocation points. One common strategy is to use Gauss-Legendre quadrature, which we will also use for numerical integration in subsequent chapters. Since we need to generate two equations, we will use two integration points as provided by quadrature rules:

$$x_i = \pm \frac{1}{\sqrt{3}} \tag{6.42}$$

Quadrature points are always specified within a normalized range $[-1, +1]$ requiring that we transform the points to the geometry of our beam $[0, L]$:

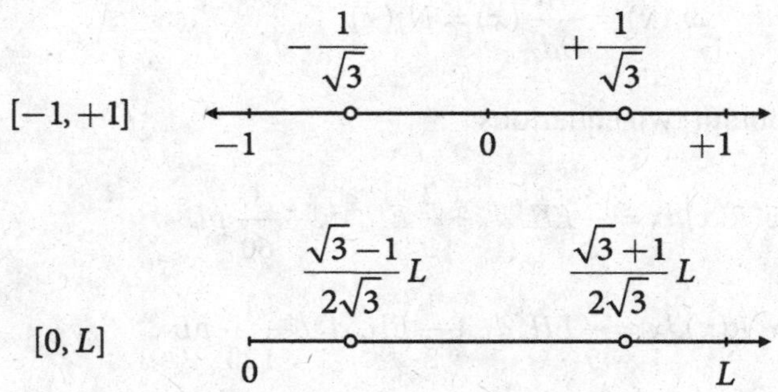

$$[-1, +1]$$

$$-\frac{1}{\sqrt{3}} \qquad +\frac{1}{\sqrt{3}}$$

$$-1 \qquad 0 \qquad +1$$

$$[0, L]$$

$$\frac{\sqrt{3}-1}{2\sqrt{3}}L \qquad \frac{\sqrt{3}+1}{2\sqrt{3}}L$$

$$0 \qquad L$$

Figure 6.6. Coordinate transformation for Gaussian quadrature.

Substituting these two points into our residual provides us with two equations

$$0 = 6EId_3 \left(\frac{\sqrt{3}-1}{2\sqrt{3}} L \right) + 2EId_2 + \frac{p}{2} \left(L - \frac{\sqrt{3}-1}{2\sqrt{3}} L \right)^2$$

$$0 = 6EId_3 \left(\frac{\sqrt{3}+1}{2\sqrt{3}} L \right) + 2EId_2 + \frac{p}{2} \left(L - \frac{\sqrt{3}+1}{2\sqrt{3}} L \right)^2 \tag{6.43}$$

Solving this set of equations generates the same dofs found by our VP:

$$d_3 = \frac{pL}{12EI}; \quad d_2 = -\frac{5pL^2}{24EI} \tag{6.44}$$

6.5 Evaluating the Approximations

Using the exact solution as a benchmark, we evaluate the approximate solutions at the boundaries and within the domain. Because we are investigating a differential equation, we must assess not only the behavior of the function, but also the relevant derivatives. For our example, we include the vertical displacement, $\tilde{v}(x)$, the rotation, $\tilde{\theta}(x)$, and the moment, $\tilde{M}(x)$, in our comparison.

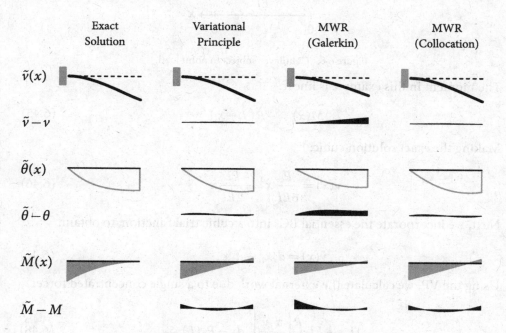

Figure 6.7. Comparison of approximate solutions.

All of the approximations performed well; they exhibit the same general behavior and magnitude as the exact solution. Because essential BCs were incorporated into the trial function, all three approximations satisfy the essential BCs. Due to our choice of cubic trial function, none of the approximations were able to satisfy the natural BCs represented by the moment diagram at more than two points; all approximate solutions met this maximum number of points.

Even though each solution is optimal for its approximation technique, there is no single best approximation strategy; ultimately, each method we have investigated is just one strategy that works well for a specific class of problems.

6.6 Approaching the Exact Solution

As a general rule, if the trial function is capable of reproducing the exact solution, it will. To demonstrate this principle, consider a cantilever subject to a point load:

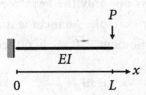

Figure 6.8. Cantilever subject to point load.

The moment in this example is linear:

$$M(x) = -P\left(L - x\right) \tag{6.45}$$

Making the exact solution cubic:

$$v(x) = \frac{P}{6EI} x^3 - \frac{PL}{2EI} x^2 \tag{6.46}$$

Next, we incorporate the essential BCs into a cubic trial function, to obtain:

$$\tilde{v}(x) = d_3 x^3 + d_2 x^2 \tag{6.47}$$

Using the VP, we calculate the external work due to a single concentrated force:

$$\Pi = \tfrac{1}{2} EI \int_L \left(\frac{d^2 \tilde{v}}{dx^2}\right)^2 dx - P\tilde{v}(L) \tag{6.48}$$

Integrating, we get the polynomial expression:

$$\Pi = 6EIL^3 d_3^2 + 6EIL^2 d_3 d_2 + 2EILd_2^2 - PL^3 d_3 - PL^2 d_2 \tag{6.49}$$

Once again, we find two equations:

$$\frac{\partial \Pi}{\partial d_2} = 0 = 6EIL^2 d_3 + 4EILd_2 - PL^2$$

$$\frac{\partial \Pi}{\partial d_3} = 0 = 12EIL^3 d_3 + 6EIL^2 d_2 - PL^3 \tag{6.50}$$

Solving these two equations leads us to an approximate solution which matches the exact solution:

$$v(x) = \frac{P}{6EI} x^3 - \frac{PL}{2EI} x^2 \qquad (6.51)$$

If we use the MWR, we arrive at similar results. First, we define the residual:

$$R(x) = 6EId_3 x + 2EId_2 + P(L - x) \qquad (6.52)$$

If we use the Galerkin method, our two weighted integrations produce the following two equations:

$$0 = \int_L x^2 R(x)dx = \frac{3}{2} EIL^4 d_3 + \frac{2}{3} EIL^3 d_2 + \frac{1}{12} PL^4$$

$$0 = \int_L x^3 R(x)dx = \frac{6}{5} EIL^5 d_3 + \frac{1}{2} EIL^4 d_2 + \frac{1}{20} PL^5 \qquad (6.53)$$

The approximate solution once again matches the exact solution:

$$v(x) = \frac{P}{6EI} x^3 - \frac{PL}{2EI} x^2 \qquad (6.54)$$

If we employ the collocation method using two quadrature points, we obtain the following set of equations:

$$0 = 6EId_3 \left(\frac{\sqrt{3} - 1}{2\sqrt{3}} L \right) + 2EId_2 + P \left(L - \frac{\sqrt{3} - 1}{2\sqrt{3}} L \right)$$

$$0 = 6EId_3 \left(\frac{\sqrt{3} + 1}{2\sqrt{3}} L \right) + 2EId_2 + P \left(L - \frac{\sqrt{3} + 1}{2\sqrt{3}} L \right) \qquad (6.55)$$

The approximate solution again reproduces the exact solution:

$$v(x) = \frac{P}{6EI} x^3 - \frac{PL}{2EI} x^2 \qquad (6.56)$$

6.7 Setting up a Framework for FEM

The MWR is the more powerful of the two approximation techniques we have examined because it provides a systematic method for turning the strong form of any BVP into an equivalent weak form. By comparison, VPs are available only for self-adjoint differential equations (a condition which incidentally guarantees the symmetry of our stiffness matrices). Fortunately, both phenomena that we will investigate (linear elasticity and stead-state heat) have readily available VPs. The practical advantage of the VP is that it involves less integration than the MWR, thereby reducing the complexity of the mathematics involved. Though both the MWR and the VP can be used to derive FEM elements, we will use the VP exclusively in this text because it allows us to obtain stiffness formulations efficiently.

Our next task is to adapt the VP into a form equivalent to the stiffness equations we saw in MSA. We have already identified familiar terms (i.e., natural and essential BCs, systems of equations, and dofs), but we still need to devise a methodical approach, starting by defining the potential. For 1D elasticity, the potential is defined by the potential energy:

$$\Pi = \int_\Omega \frac{E}{2}\left(\frac{\partial u}{\partial x}\right)^2 dV - \int_\Omega ufdV - \int_\Gamma u\bar{\sigma}dA \tag{6.57}$$

For steady-state heat, the potential in 1D is mathematically obtained, but maintains a familiar form:

$$\Pi = \int_\Omega \frac{k}{2}\left(\frac{\partial T}{\partial x}\right)^2 dV - \int_\Omega TQdV - \int_\Gamma T\bar{q}dA \tag{6.58}$$

In higher dimensions, the heat potential remains parabolic while the static elasticity equation becomes elliptic. Nonetheless, the potential for both elasticity and heat always maintains components corresponding to internal effects, U, external effects due to body forces, W_Ω, and external effects due to surface forces, W_Γ:

$$\Pi = U - W_\Omega - W_\Gamma \tag{6.59}$$

Of the three components, the first two occur within the domain, and thus can be expressed jointly as an integral over the domain volume. The third term occurs on the boundary of the domain and thus can be expressed as a surface integral over the domain. The integrand in each expression, g and h, can be generally expressed as operators of the function and its higher derivatives:

$$U - W_\Omega = \int_\Omega g\left(u, \frac{\partial u}{\partial x}, ...\right) dV \; ; \quad W_\Gamma = \int_\Gamma h\left(u, \frac{\partial u}{\partial x}, ...\right) dA \qquad (6.60)$$

Because both expressions are definite integrals, the potential will generate a **functional**, $\Pi(u, \partial u/\partial x, ...)$, which is a function of functions, distinct from a primary function, $u(x)$, which is a function of variables.

The only independent variables in a trial function are the unknown dofs:

$$\tilde{u}(d_1, d_2, d_3...) = \sum_{neq} d_j N_j \qquad (6.61)$$

Upon introducing the trial function, the potential becomes a function of the dofs:

$$\Pi = \Pi(\tilde{u}, \partial \tilde{u}/\partial x, ...) = \Pi(d_1, d_2, d_3...) \qquad (6.62)$$

Employing the Variational Principle, we set each of the partial derivatives to zero:

$$\frac{\partial \Pi}{\partial d_i} = 0 \qquad (6.63)$$

This minimization produces a set of equations (one for each dof) that we can reformulate as the familiar stiffness equation:

$$\frac{\partial \Pi}{\partial d_i} = [K]\{d\} - \{F\} = 0 \qquad (6.64)$$

The matrix components will relate to our internal and external effects:

$$[K]\{d\} = \frac{\partial U}{\partial d_i} \; ; \quad \{F\} = \frac{\partial (W_\Omega + W_\Gamma)}{\partial d_i} \qquad (6.65)$$

We cannot proceed further without delving into mathematical specifics. In subsequent chapters, we will demonstrate how these steps unfold for steady-state heat and linear elasticity. Our implementation of FEM will differ from MSA in element derivation; assembly operations, BCs, and dofs solution remain identical.

Chapter 7

Steady-State Heat Conduction

We begin our investigation of FEM with steady-state heat conduction, a common engineering problem, which allows us to demonstrate the fundamentals of FEM without the complexities inherent to linear elasticity. Whereas the primary unknown in elasticity is the 3D vector field of displacement, heat conduction is concerned with the scalar field of temperature. Heat conduction problems are limited to one temperature dof per node, thus greatly reducing the complexity of the mathematics involved in our stiffness derivations.

In this chapter, we begin by presenting the strong form of the heat conduction BVP. Next, we use the Variational Principle to derive the basic formulation of the stiffness equation, which we then use to formulate rod, triangle, and quad stiffness matrices for 1D and 2D heat conduction problems. To aid in these derivations, we introduce several new concepts including shape functions, mapping, and numerical integration. Following the theoretical derivations, we present code for the element function. We conclude with an example analysis.

7.1 Strong Form of the BVP

Heat is a form of energy, a quantity measured in Joules (J). **Heat conduction** describes the flow of heat over time and is thus measured in Watts (J/s). While a steady-state, stationary, or static condition is never fully achievable, it can be a reasonable simplification of real-life scenarios. A steady-state heat condition is approached when a physical body is subject to invariable fluxes over a long duration of time. Permitting such simplifications, we derive the governing differential equations and boundary conditions of the steady-state heat BVP.

131

7.1.1 Governing Differential Equations

Steady-state heat conduction is governed by constitutive and conservative laws. The **constitutive** law for heat conduction was derived by Joseph Fourier in 1822 and is aptly termed **Fourier's law**:

$$q = -k \frac{dT}{dx} \tag{7.1}$$

In this formulation, q is the **heat flux**, k is the **thermal conductivity**, and dT/dx is the **temperature gradient**. Heat flux is the amount of energy that flows through a unit of area over time and is typically measured in W/m² (Watts per square meter). Thermal conductivity is a material property defined by the quantity of thermal energy that can pass through a unit of thickness and is typically measured in W/m°C (Watts per meter-Celsius). High thermal conductivity indicates that the material is a poor insulator and transmits thermal energy easily. In this text, we use Celsius (°C) to measure temperature.

In 3D, Fourier's law is expressed as a set of three directional fluxes stored in the **heat flux vector**, $\{q\}$, and expressed as the **gradient**, ∇, of the temperature.

$$\{q_x \; q_y \; q_z\} = \{q\} = -k\nabla T = -k \left\{ \frac{\partial T}{\partial x} \; \frac{\partial T}{\partial y} \; \frac{\partial T}{\partial z} \right\} \tag{7.2}$$

This expression is simplified for an isotropic condition where the thermal conductivity is equal in all directions ($k_x = k_y = k_z = k$).

Although steady-state heat conduction is not in equilibrium, it needs to satisfy the law of energy conservation. To illustrate heat conduction in a solid, consider an infinitesimal cube with sides, $dx = dy = dz$, faces, dA, and volume, dV :

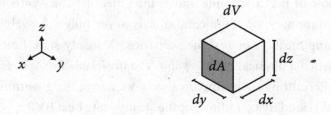

Figure 7.1. Infinitesimal cube in isometric projection.

Under steady-state heat conduction, the cube is affected by both heat sources and heat fluxes. **Volumetric heat source**, Q, is measured in W/m³ and assumed to be constant throughout the volume. Although the heat flux is assumed to change slightly as it passes through the volume, the rate of change in heat flux, $\partial q_x / \partial x$, is assumed to remain constant. Taking the center of the cube as a reference point, we present the heat source and heat flux contributions visually:

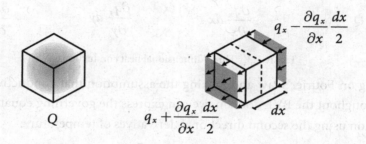

Figure 7.2. One-dimensional heat conduction.

Steady-state heat is subject to the law of conservation of energy. In order to balance out the two contributions, we must obtain consistent units; we multiply the volumetric heat density by the infinitesimal volume, dV, and multiply the ingoing and outgoing fluxes by the infinitesimal areas, dA. Employing the sign convention that heat entering the cube is treated as positive, we sum the two weighted contributions to establish energy conservation:

$$0 = (Q)dV + \left(q_x - \frac{\partial q_x}{\partial x} \frac{dx}{2} \right) dA - \left(q_x + \frac{\partial q_x}{\partial x} \frac{dx}{2} \right) dA \qquad (7.3)$$

Gathering terms and using the equivalence $dV = dxdA$, we arrive at the 1D equation of energy conservation for heat conduction:

$$0 = Q - \frac{\partial q_x}{\partial x} \qquad (7.4)$$

We can easily expand this formulation to 3D:

$$0 = Q - \left(\frac{\partial q_x}{\partial x} + \frac{\partial q_y}{\partial y} + \frac{\partial q_z}{\partial z} \right) \qquad (7.5)$$

The 3D heat sources and fluxes are visualized below:

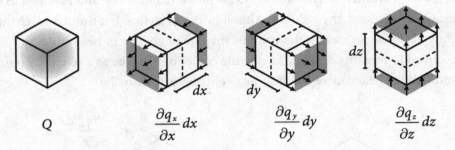

$$Q \qquad \frac{\partial q_x}{\partial x}dx \qquad \frac{\partial q_y}{\partial y}dy \qquad \frac{\partial q_z}{\partial z}dz$$

Figure 7.3.   Three-dimensional heat conduction.

By calling on Fourier's law and making the assumption that conductivity is constant throughout the BVP domain, we can express the governing equation of heat conduction using the second directional derivatives of temperature:

$$0 = Q + k\left(\frac{\partial^2 T_x}{\partial x^2} + \frac{\partial^2 T_y}{\partial y^2} + \frac{\partial^2 T_z}{\partial z^2}\right) \qquad (7.6)$$

7.1.2 Boundary Conditions

Having derived the governing differential equations over the domain of the BVP, we next need to define the BCs. The essential BCs are defined as **prescribed temperatures**, T_s, along the boundary while the natural BCs are defined by the **applied flux**, $\overline{q}$, applied to the system:

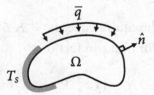

Figure 7.4.   BCs of heat-conduction BVP.

While the total flux at any point along the boundary may be oriented in any direction, we only want to capture the component that enters the system. Thus, we only use the flux that is parallel to the **bounding surface normal**, $\hat{n}$. Because this normal points outward, opposite the direction of positive heat flux entering the system, the applied flux is the negative of the dot product of the two vectors:

$$\bar{q} = -\left(n_x q_x + n_y q_y + n_z q_z\right) \tag{7.7}$$

We prioritize the essential BC if both natural and essential BCs are specified at any point on a boundary; if no BC is specified, a null applied flux is assumed.

7.2 Derivation of the Stiffness Equation Using the VP

In this section, we will follow the general steps demonstrated in the previous chapter to derive the explicit stiffness equation for heat. First, we will establish the heat potential and introduce the trial function for temperature. Then, we will prepare the partial derivatives of the trial function and its directional derivatives, which we will use to find an expression for the minimized potential. From this minimized potential, we will extract the components of the stiffness equation.

For elasticity, we were able to utilize the physical quantity of potential energy to develop the VP formulation. Since there is no analogous physical quantity for heat, we must rely on the mathematically obtained potential shown below:

$$\Pi = \frac{k}{2} \int_{\Omega} \left(\frac{\partial T}{\partial x}\right)^2 + \left(\frac{\partial T}{\partial y}\right)^2 + \left(\frac{\partial T}{\partial z}\right)^2 \, dV - \int_{\Omega} TQ dV - \int_{\Gamma} T\bar{q} dA \tag{7.8}$$

Though it is uncomfortable to accept this equation as given, its derivation relies on advanced topics in the calculus of variations that are outside the scope of this book. The heat potential is still analogous to the elastic potential:

$$\Pi = U - W_{\Omega} - W_{\Gamma} \tag{7.9}$$

The potential components, U, W_{Ω}, and W_{Γ}, relate respectively to the internal heat conduction, internal heat source, and external flux entering the system.

The trial function for temperature is defined as the sum product of temperature dofs, d_Q, and normalized basis functions, N_Q:

$$\tilde{T}(x, y, z) = \sum_{Q=1:neq} d_Q N_Q(x, y, z) \tag{7.10}$$

The subscript Q identifies the global dof; as we will see later, this choice allows us to remain consistent with the subscript convention established in Chapter 3.

The potential also contains directional derivatives of temperature, which reduce to a weighted sum of dofs and directional derivatives of basis functions:

$$\frac{\partial \tilde{T}}{\partial x} = \sum_Q d_Q \frac{\partial N_Q}{\partial x} \tag{7.11}$$

Using temperature and its directional derivatives, we can state the components of the potential equation:

$$U = \frac{k}{2} \int_\Omega \left(\sum_Q d_Q \frac{\partial N_Q}{\partial x} \right)^2 + \left(\sum_Q d_Q \frac{\partial N_Q}{\partial y} \right)^2 + \left(\sum_Q d_Q \frac{\partial N_Q}{\partial z} \right)^2 dV \tag{7.12}$$

$$W_\Omega + W_\Gamma = \int_\Omega Q \left(\sum_Q d_Q N_Q \right) dV + \int_\Gamma \bar{q} \left(\sum_Q d_Q N_Q \right) dA \tag{7.13}$$

Guided by the VP, we minimize the potential by taking its partial derivatives with respect to each dof. We use the subscript P to distinguish the partial derivative indices from the trial function subscripts:

$$\frac{\partial \Pi}{\partial d_P} = \frac{\partial U}{\partial d_P} - \frac{\partial W_\Omega}{\partial d_P} - \frac{\partial W_\Gamma}{\partial d_P} = 0 \tag{7.14}$$

Since the potential includes the trial function and the squares of its directional derivatives, we need to find their partial derivatives. The partial derivative of the trial function, $\tilde{T}$, with respect to d_P reduces to the associated shape function:

$$\frac{\partial \tilde{T}}{\partial d_P} = \frac{\partial}{\partial d_P} \left(\sum_Q d_Q N_Q \right) = \sum_Q \frac{\partial d_Q}{\partial d_P} N_Q = N_P \tag{7.15}$$

The partial derivative of the directional derivative of the trial function, $\partial \tilde{T}/\partial x$, is the directional derivative of the associated shape function:

$$\frac{\partial}{\partial d_P} \left(\frac{\partial \tilde{T}}{\partial x} \right) = \frac{\partial N_P}{\partial x} \tag{7.16}$$

The partial derivative of the square of the trial function directional derivative is a bit trickier to find:

$$\frac{\partial}{\partial d_P}\left(\frac{\partial \tilde{T}}{\partial x}\right)^2 = 2\left(\frac{\partial \tilde{T}}{\partial x}\right)\frac{\partial}{\partial d_P}\left(\frac{\partial \tilde{T}}{\partial x}\right) = 2\left(\sum_Q d_Q \frac{\partial N_Q}{\partial x}\right)\frac{\partial N_P}{\partial x} \qquad (7.17)$$

Using these relationships, we express the partial derivatives of the components of the potential equation:

$$\frac{\partial U}{\partial d_P} = k \int_\Omega \left(\sum_Q \left(\frac{\partial N_P}{\partial x}\frac{\partial N_Q}{\partial x} + \frac{\partial N_P}{\partial y}\frac{\partial N_Q}{\partial y} + \frac{\partial N_P}{\partial z}\frac{\partial N_Q}{\partial z} \right) d_Q \right) dV \qquad (7.18)$$

$$\frac{\partial W_\Omega}{\partial d_P} + \frac{\partial W_\Gamma}{\partial d_P} = \int_\Omega Q N_P dV + \int_\Gamma \overline{q} N_P dA \qquad (7.19)$$

The partial derivatives of the potential actually supply all of the components of the matrix stiffness equation:

$$\frac{\partial \Pi}{\partial d_P} = 0 = [K]\{d\} - \{F\} \qquad (7.20)$$

The entries of the stiffness matrix are taken from the internal conduction term:

$$K_{PQ} = k \int_\Omega \left(\frac{\partial N_P}{\partial x}\frac{\partial N_Q}{\partial x} + \frac{\partial N_P}{\partial y}\frac{\partial N_Q}{\partial y} + \frac{\partial N_P}{\partial z}\frac{\partial N_Q}{\partial z} \right) dV \qquad (7.21)$$

The displacement vector is directly populated by the dofs:

$$d_Q = d_Q \qquad (7.22)$$

The force vector is assembled out of contributions from the internal heat intensity and external heat flux terms:

$$F_P = \int_\Omega Q N_P dV + \int_\Gamma \overline{q} N_P dA \qquad (7.23)$$

The force vector is assembled by iterating through the index P; the displacement vector is assembled by iterating through the index Q; the stiffness matrix is assembled by iterating through both indices P and Q.

7.3 Basis and Shape Functions

The trial function discretizes the solution space into pairs of dofs and basis functions. We extend this discretization to the geometry of the problem by applying a mesh to the domain of the BVP:

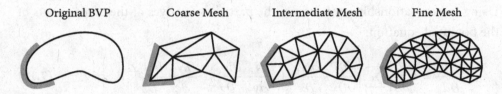

| Original BVP | Coarse Mesh | Intermediate Mesh | Fine Mesh |

Figure 7.5.   Meshing alternatives using triangles.

A **mesh** defines both nodal dofs and compatible basis functions. Though meshes cannot necessarily replicate the geometry of a BVP, particularly curved boundaries, we must remember that we are working with approximate solutions where approximate geometric meshes are permissible. A general principle of FEM is that the approximation accuracy should improve with mesh resolution.

An approximate solution over a mesh is easily formed using basis **hat functions** coupled with nodal dofs. Hat functions have three primary characteristics: first, they are equal to one at their associated node (thus making the approximate solution equal to the dof at this point); second, they are nonzero only within the elements adjacent to the primary node; and third, they reduce to zero at the perimeter bounding adjacent elements. A single basis hat function can be constructed per the diagram below:

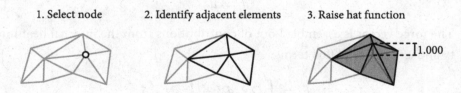

| 1. Select node | 2. Identify adjacent elements | 3. Raise hat function |

Figure 7.6.   Constructing a hat function.

We can apply this procedure to generate basis functions at every node of a mesh:

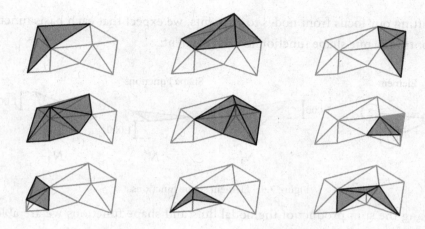

Figure 7.7. Complete set of basis hat functions in the coarse mesh.

On inspection, we note that all of the hat functions are linearly independent; it is impossible to construct any one basis function using the other functions. Furthermore, each hat function can be broken up into piecewise element contributions called **shape functions**:

Basis Function   Contributing Shape Functions

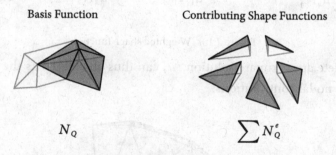

$$N_Q \hspace{5cm} \sum N_Q^e$$

Figure 7.8. Contributing shape functions to basis functions.

It is important to note that the difference between basis and shape functions is subtle: basis functions are defined over the entire BVP domain whereas shape functions are defined only within each element domain. Each basis function can thus be defined as the sum of the shape functions from adjacent elements (being careful not to double-count contributions at element boundaries):

$$N_Q(x, y, z) = \begin{cases} \sum N_Q^e & \text{if } (x, y, z) \in \Omega^e \\ N_Q^e & \text{if } (x, y, z) \in \Gamma^e \end{cases} \tag{7.24}$$

Shifting our focus from nodes to elements, we expect that each basis function will contribute one shape function to each element:

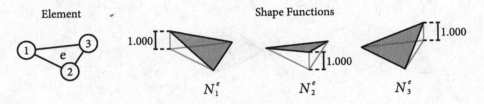

Figure 7.9.   Element shape functions.

By taking the sum product of the nodal dofs and shape functions we are able to construct the element contribution to the approximate solution, $\tilde{u}^e$:

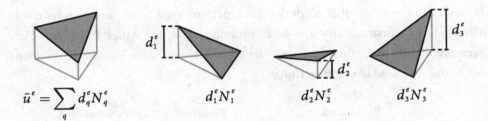

Figure 7.10.   Weighted shape functions.

The complete approximate solution, $\tilde{u}$, can thus be defined as the sum of either element or nodal contributions:

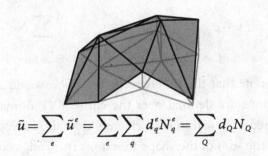

$$\tilde{u} = \sum_e \tilde{u}^e = \sum_e \sum_q d_q^e N_q^e = \sum_Q d_Q N_Q$$

Figure 7.11.   Assembling the complete approximate solution.

7.3.1 Shape Functions in the Physical Domain

Although we have significant freedom in selecting shape functions, we generally like to choose shape functions that adhere to three simple rules.

Firstly, each shape function should be equal to one at its associated node, but equal to zero at all other element nodes. Not only does this condition guarantee that we will be able to construct linearly independent basis hat functions, but it also permits direct representation of the approximate solution at nodes. We can express this rule using the **Kronecker delta**, δ_{ij}:

$$N_i^e(x_j, y_j) = \delta_{ij} = \begin{cases} 0 \text{ if } i \neq j \\ 1 \text{ if } i = j \end{cases} \tag{7.25}$$

Secondly, shape functions should be able to capture as large a range of solutions as possible. This flexibility is attained through **completeness**, which necessitates that a weighted sum of the shape functions should be able to reproduce any polynomial of a certain degree (typically the highest degree of any shape function). This requirement is critical for ensuring that the approximate solution converges to the exact solution as the mesh is refined (more elements are used). A simple way to check completeness is to show that the unweighted sum of the shape functions within the element is equal to one:

$$\sum_{i=1:nen} N_i^e = 1 \tag{7.26}$$

Lastly, we require that shape functions generate a solution that is sufficiently **smooth** or **continuous**. A function is classified as C^n continuous when its nth and all lower derivatives exist and are continuous. For instance, a C^0 function is continuous over $f(x)$, while a C^1 function is continuous over both $f(x)$ and df/dx. In most FEM applications, the approximate solution must be at least C^0 continuous. Because we use partial derivatives to generate element stiffness matrices, shape functions should also be at least C^1 continuous within an element's domain. At element boundaries, we can typically only achieve C^0 continuity since hat functions produce kinks in the approximate solution at element edges.

These guidelines permit a variety of shape functions that directly influence the complexity of the resulting elements. In general, we can use many simple elements or fewer, more complex elements. While more complex shape functions may suggest greater flexibility in capturing the modelled behavior, they are also more likely to behave in unexpected ways and achieve ill-conditioned states. When possible, the more reliable approach is to use many simple elements whose behavior is intuitive and easy to understand.

In our investigation of FEM, we will use three element geometries: a rod in 1D, a triangle in 2D, and a quad in 2D. We define these three elements in their general physical domains:

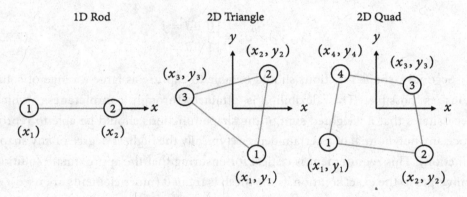

Figure 7.12.   Global element geometries.

For each of these elements, we will visualize and define the nodal shape functions. For the 1D rod, we use **linear** shape functions:

Figure 7.13.   Global rod shape functions.

These two shape functions can be defined mathematically:

$$N_1^e = \frac{x_2 - x}{x_2 - x_1}; \quad N_2^e = \frac{x - x_1}{x_2 - x_1} \tag{7.27}$$

We next need to demonstrate how these two shape functions satisfy the three requirements described at the start of this section. First, we show that each shape function is equal to one at its corresponding node and zero at the other node:

$$N_1^e(x_1) = \frac{x_2 - (x_1)}{x_2 - x_1} = 1; \quad N_1^e(x_2) = \frac{x_2 - (x_2)}{x_2 - x_1} = 0$$

$$N_2^e(x_1) = \frac{(x_1) - x_1}{x_2 - x_1} = 0; \quad N_2^e(x_2) = \frac{(x_2) - x_1}{x_2 - x_1} = 1 \qquad (7.28)$$

Next, we show that the sum of the shape functions is equal to one:

$$\sum_{i=1:2} N_i^e = \frac{x_2 - x}{x_2 - x_1} + \frac{x - x_1}{x_2 - x_1} = \frac{x_2 - x_1}{x_2 - x_1} = 1 \qquad (7.29)$$

Finally, we demonstrate that the solution space within the element is C^1 continuous by showing that the derivative of both shape functions exists:

$$\frac{dN_1^e}{dx} = \frac{1}{x_1 - x_2}; \quad \frac{dN_2^e}{dx} = \frac{1}{x_2 - x_1} \qquad (7.30)$$

If we visualize a multi-element section of a representative approximate solution, we see that while the trial function remains continuous, there are kinks at each node because the derivative is discontinuous at element boundaries.

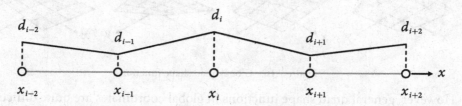

Figure 7.14. Approximate solution in 1D using linear shape functions.

Next we look at the 2D triangle, for which we use **planar** shape functions, which are easy to visualize using isometric projection:

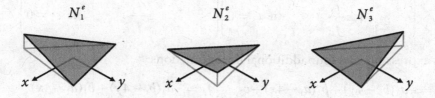

Figure 7.15. Global triangle shape functions.

These functions are slightly more complex when expressed mathematically:

$$N_1^e = \frac{x_2y_3 - x_3y_2 + (y_2 - y_3)x + (x_3 - x_2)y}{(x_2 - x_1)(y_3 - y_1) - (x_3 - x_1)(y_2 - y_1)};$$

$$N_2^e = \frac{x_3y_1 - x_1y_3 + (y_3 - y_1)x + (x_1 - x_3)y}{(x_2 - x_1)(y_3 - y_1) - (x_3 - x_1)(y_2 - y_1)};$$

$$N_3^e = \frac{x_1y_2 - x_2y_1 + (y_1 - y_2)x + (x_2 - x_1)y}{(x_2 - x_1)(y_3 - y_1) - (x_3 - x_1)(y_2 - y_1)} \tag{7.31}$$

The reader is encouraged to demonstrate that these shape functions satisfy all three of the conditions we have established.

The 2D quad shape functions are described by warped **hyperbolic** surfaces, which are again sufficiently easy to visualize:

N_1^e N_2^e N_3^e N_4^e

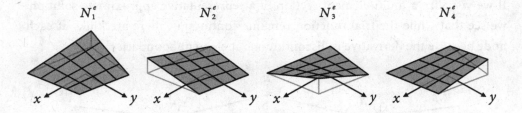

Figure 7.16. Global quad shape functions.

However, general quad shape functions in global coordinates are quite difficult to express mathematically:

$$N_i^e = \left(\begin{array}{cc} \dfrac{1}{2} \pm \dfrac{f_1 - \sqrt{f_1^2 - 4c_1f_2}}{4c_2} & \text{if } c_1 \neq 0 \\[2ex] 1 \pm \dfrac{f_2}{f_1} & \text{if } c_1 = 0 \end{array} \right| \left. \begin{array}{cc} \dfrac{1}{2} \pm \dfrac{f_3 - \sqrt{f_3^2 - 4c_2f_4}}{4c_2} & \text{if } c_2 \neq 0 \\[2ex] 1 \pm \dfrac{f_4}{f_3} & \text{if } c_2 = 0 \end{array} \right) \tag{7.32}$$

This expression uses four additional linear functions:

$$f_1 = -a_4(b_1 - 4y) + b_4(a_1 - 4x) + c_3; \quad f_2 = -a_3(b_1 - 4y) + b_3(a_1 - 4x);$$

$$f_3 = +a_4(b_1 - 4y) - b_4(a_1 - 4x) + c_3; \quad f_4 = +a_2(b_1 - 4y) - b_2(a_1 - 4x) \tag{7.33}$$

The coefficients to these equations are supplied below:

$$c_1 = a_2 b_4 - a_4 b_2; \quad c_2 = a_4 b_3 - a_3 b_4; \quad c_3 = a_2 b_3 - a_3 b_2 \tag{7.34}$$

For ease of expression, we also define several coordinate permutations:

$$
\begin{aligned}
a_1 &= +x_1 + x_2 + x_3 + x_4 & b_1 &= +y_1 + y_2 + y_3 + y_4 \\
a_2 &= -x_1 + x_2 + x_3 - x_4 & b_2 &= -y_1 + y_2 + y_3 - y_4 \\
a_3 &= -x_1 - x_2 + x_3 + x_4 & b_3 &= -y_1 - y_2 + y_3 + y_4 \\
a_4 &= +x_1 - x_2 + x_3 - x_4 & b_4 &= +y_1 - y_2 + y_3 - y_4
\end{aligned}
\tag{7.35}
$$

As apparent by the increasing complexity of these shape functions, working in global coordinates becomes very laborious very quickly. We are thus motivated to develop convenient forms of the shape functions in a local coordinate system.

7.3.2 Shape Functions in the Parent Domain

As we did for the truss, beam, and frame elements, we want to define a local coordinate system for elements. In FEM, we need to do more than rotate elements; instead, we need to stretch and distort elements between coordinate systems. As a result, we use the terms **parent** and **physical domains** to refer to the local and global coordinate systems when talking about FEM element geometries. The parent domain is defined by a set of axes labelled using the Greek letters, ξ - "xi", η - "eta", and ζ - "zeta", corresponding to the global axes, x, y, and z. In order to create simple shape functions, we choose simple local element geometries:

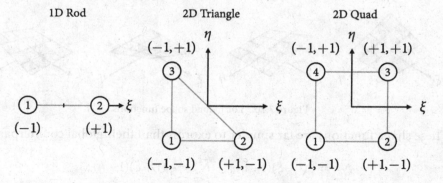

Figure 7.17. Parent element geometries.

The **local shape functions**, $\bar{N}_i^e$, for the rod are again linear and easy to visualize:

Figure 7.18.   Local rod shape functions.

They are even easier to define mathematically:

$$\bar{N}_1^e = \tfrac{1}{2}(1-\xi); \quad \bar{N}_2^e = \tfrac{1}{2}(1+\xi) \tag{7.36}$$

The triangle shape functions remain planar and are also easy to visualize:

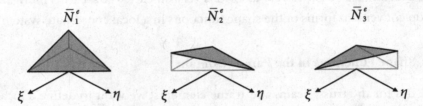

Figure 7.19.   Local triangle shape functions.

They are again simple to define mathematically:

$$\bar{N}_1^e = \tfrac{1}{2}(-\xi-\eta); \quad \bar{N}_2^e = \tfrac{1}{2}(1+\xi); \quad \bar{N}_3^e = \tfrac{1}{2}(1+\eta) \tag{7.37}$$

The quad shape functions are hyperbolic paraboloids:

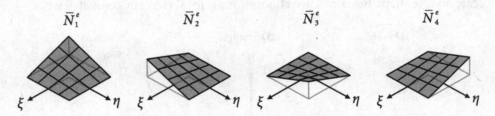

Figure 7.20.   Local quad shape functions.

These shape functions are far simpler to express than their global counterparts:

$$\bar{N}_1^e = \tfrac{1}{4}(1-\xi)(1-\eta); \quad \bar{N}_2^e = \tfrac{1}{4}(1+\xi)(1-\eta);$$

$$\bar{N}_3^e = \tfrac{1}{4}(1+\xi)(1+\eta); \quad \bar{N}_4^e = \tfrac{1}{4}(1-\xi)(1+\eta) \tag{7.38}$$

The reader should verify that all three sets of shape functions satisfy our shape function requirements. It is worth noting that the shape functions for all three elements are compatible, meaning that combinations of these elements will be C^0 continuous across element boundaries.

7.3.3 Mapping between Domains

Having defined shape functions in the parent domain, we next need to establish a **mapping** strategy between the parent and physical domains. **Forward mapping** (going from parent to physical domain) is simple; we just need to perform the sum product of the shape functions and their associated nodal coordinates:

$$x(\xi,\eta) = \sum_{i=1:nen} x_i \bar{N}_i^e(\xi,\eta); \quad y(\xi,\eta) = \sum_{i=1:nen} y_i \bar{N}_i^e(\xi,\eta) \tag{7.39}$$

In 1D, these relationships only relate ξ to x:

$$x(\xi) = \sum_{i=1:nen} x_i \bar{N}_i^e(\xi) \tag{7.40}$$

Because the shape functions are proven to be C^0 continuous across element boundaries, every point in the physical domain relates to a point in the parent domain and vice-versa; there are no gaps in forward or backward mapping.

Backward mapping (going from physical to parent domain) is more difficult and approaches the algebraic complexity we encountered in establishing the global shape functions. For the 1D rod, the local coordinates are found using the linear relationship:

$$\xi = \frac{2x - x_1 - x_2}{x_2 - x_1} \tag{7.41}$$

For the 2D triangle, the local coordinates are found using a planar relationship:

$$\xi = \frac{(y_3 - y_1)(2x - x_2) + (x_1 - x_3)(2y - y_2) - x_1 y_3 + x_3 y_1}{x_1(y_2 - y_3) + x_2(y_3 - y_1) + x_3(y_1 - y_2)} \tag{7.42}$$

$$\eta = \frac{(y_1 - y_2)(2x - x_3) + (x_2 - x_1)(2y - y_3) + x_1 y_2 - x_2 y_1}{x_1(y_2 - y_3) + x_2(y_3 - y_1) + x_3(y_1 - y_2)} \tag{7.43}$$

For the 2D quad, the local coordinates are expressed using the coefficients from the global shape functions:

$$\xi = \begin{cases} -\dfrac{f_1 - \sqrt{f_1^2 - 4c_1 f_2}}{2c_1} & \text{if } c_1 \neq 0 \\[4mm] -\dfrac{f_2}{f_1} & \text{if } c_1 = 0 \end{cases} \tag{7.44}$$

$$\eta = \begin{cases} -\dfrac{f_3 - \sqrt{f_3^2 - 4c_2 f_4}}{2c_2} & \text{if } c_2 \neq 0 \\[4mm] -\dfrac{f_4}{f_3} & \text{if } c_2 = 0 \end{cases} \tag{7.45}$$

In order to generate element stiffness matrices, we need to be able to integrate shape functions. Due to the simplicity of local shape functions, it is frequently easier to perform the integration in the parent domain following a change of variables. In 1D, the change of variable is accomplished easily:

$$\int_{x_1}^{x_2} f(x)dx = \int_{\xi_1}^{\xi_2} \frac{dx}{d\xi} f(x(\xi))d\xi \tag{7.46}$$

In 2D, we have two variables, but the integration follows the same structure:

$$\int_{y_1}^{y_2} \int_{x_1}^{x_2} f(x, y)dxdy = \int_{\eta_1}^{\eta_2} \int_{\xi_1}^{\xi_2} |J| f(x(\xi, \eta), y(\xi, \eta))d\xi d\eta \tag{7.47}$$

The coefficient $|J|$ is the **determinant** of the **Jacobian**, $[J]$, which is defined as a matrix storing the total directional derivatives of the global coordinates taken with respect to the local coordinates:

$$[J] = \begin{vmatrix} \dfrac{\partial x}{\partial \xi} & \dfrac{\partial y}{\partial \xi} \\[3mm] \dfrac{\partial x}{\partial \eta} & \dfrac{\partial y}{\partial \eta} \end{vmatrix}; \quad |J| = \frac{\partial x}{\partial \xi}\frac{\partial y}{\partial \eta} - \frac{\partial y}{\partial \xi}\frac{\partial x}{\partial \eta} \tag{7.48}$$

7.3.4 Element Stiffness Equation

Having established element shape functions, we are prepared to develop element stiffness equations. We start by defining the BVP potential as a sum of element contributions:

$$\Pi^G = \sum_{\forall e} \Pi^e \tag{7.49}$$

Similarly, we assemble the partial derivatives of the potential element-by-element:

$$\frac{\partial \Pi^G}{\partial d_P} = \sum_{\forall e} \frac{\partial \Pi^e}{\partial d_P} \tag{7.50}$$

The global stiffnesses, displacements, and forces are assembled from element contributions just as we established in MSA:

$$K_{PQ}^G = \sum K_{PQ}^e ; \quad d_Q^G = d_Q^e ; \quad F_P^G = \sum F_P^e \tag{7.51}$$

The element stiffness equation thus takes the familiar form:

$$\left\{ F^e \right\} = \left[K^e \right] \left\{ d^e \right\} \tag{7.52}$$

The components of this element stiffness equation parallel the original global stiffness expression. The element stiffness is defined as an integral of the partial derivatives of the element contribution to the potential:

$$K_{pq}^e = k \int_{\Omega^e} \left(\frac{\partial N_p^e}{\partial x} \frac{\partial N_q^e}{\partial x} + \frac{\partial N_p^e}{\partial y} \frac{\partial N_q^e}{\partial y} + \frac{\partial N_p^e}{\partial z} \frac{\partial N_q^e}{\partial z} \right) dV \tag{7.53}$$

The change of subscripts from the global uppercase numbering (P, Q) to the element lowercase numbering (p, q) means that we only need to iterate through the element dofs in order to complete the element stiffness expression. To express the full element stiffness matrix more conveniently, we gather the terms inside the integral into a set of matrices:

$$\left[K^e \right] = \int_{\Omega^e} \left[B \right]^T \left[D \right] \left[B \right] dV \tag{7.54}$$

The **B-matrix**, $[B]$, stores the partial derivatives of the shape functions:

$$[B] = \begin{bmatrix} \dfrac{\partial N_1^e}{\partial x} & \dfrac{\partial N_2^e}{\partial x} & \cdots & \dfrac{\partial N_{nen}^e}{\partial x} \\[2ex] \left(\dfrac{\partial N_1^e}{\partial y}\right) & \left(\dfrac{\partial N_2^e}{\partial y}\right) & \cdots & \left(\dfrac{\partial N_{nen}^e}{\partial y}\right) \\[2ex] \left(\dfrac{\partial N_1^e}{\partial z}\right) & \left(\dfrac{\partial N_2^e}{\partial z}\right) & \cdots & \left(\dfrac{\partial N_{nen}^e}{\partial z}\right) \end{bmatrix} \tag{7.55}$$

The **constitutive matrix**, $[D]$, stores the directional thermal conductivities. In an anisotropic continuum, the diagonal values in this matrix will differ (k_x, k_y, k_z), but for an isotropic medium, the constitutive matrix is consistent:

$$[D] = k[I] = \begin{bmatrix} k & 0 & 0 \\ 0 & (k) & 0 \\ 0 & 0 & (k) \end{bmatrix} \tag{7.56}$$

To complete the stiffness equation, we must also define the element displacement and force vectors. The entries of the element displacement vector are trivially found:

$$d_q^e = d_q^e \tag{7.57}$$

The entries in the applied force element vector are composed of contributions from internal heat densities and any applied external heat fluxes acting on the element:

$$F_p^e = \int_{\Omega^e} Q N_p \, dV + \int_{\Gamma^e} \overline{q} N_p \, dA \tag{7.58}$$

Having established a general formulation for element stiffnesses, we allocate the subsequent sections to the derivation of specific stiffness expressions for the 1D rod, 2D triangle, and 2D quad elements.

7.4 1D Rod Element

In 1D, we assume that the temperature gradient is nonzero only along the element axis (i.e., $\partial T/\partial y = \partial T/\partial z = 0$). The potential thus reduces to:

$$\Pi = \frac{k}{2} \int_{\Omega} \left(\frac{\partial T}{\partial x} \right)^2 dV - \int_{\Omega} QT dV - \int_{\Gamma} \bar{q} T dA \qquad (7.59)$$

The two 1D rod shape functions are easy to express in global coordinates:

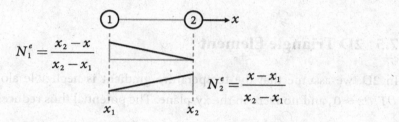

$$N_1^e = \frac{x_2 - x}{x_2 - x_1} \qquad\qquad N_2^e = \frac{x - x_1}{x_2 - x_1}$$

Figure 7.21. 1D rod geometry and shape functions in the physical domain.

Due to the simplicity of the global shape functions, we derive the element stiffness in global coordinates. In 1D, the element stiffness takes the following form:

$$K_{pq}^e = k \int_{\Omega^e} \left(\frac{\partial N_p}{\partial x} \frac{\partial N_q}{\partial x} \right) dV ; \quad p, q \in \left[1:2 \right] \qquad (7.60)$$

If we assume that the cross-sectional area, A, is constant, the element stiffness matrix can be reduced to an integral over the element length:

$$\left[K^e \right] = A \int_L \left[B \right]^T \left[D \right] \left[B \right] dx \qquad (7.61)$$

With two shape functions in one dimension, the rod B-matrix is just 1×2:

$$\left[B \right] = \left[\frac{\partial N_1^e}{\partial x} \quad \frac{\partial N_2^e}{\partial x} \right] = \frac{1}{L} \left[-1 \ +1 \right] \qquad (7.62)$$

The constitutive matrix is an even simpler 1×1 matrix:

$$\left[D \right] = \left[k \right] \qquad (7.63)$$

Since both matrices are constant, they can be removed from the integrand:

$$[K^e] = A[B]^T[D][B]\int_L dx = A[B]^T[D][B]L \tag{7.64}$$

Upon introducing both the B and constitutive matrices into this equation, we find the element stiffness matrix for the 1D rod:

$$[K^e] = A\left(\frac{1}{L}\begin{bmatrix} -1 \\ +1 \end{bmatrix}\right)([k])\left(\frac{1}{L}[-1 \ +1]\right)L = \frac{kA}{L}\begin{bmatrix} +1 & -1 \\ -1 & +1 \end{bmatrix} \tag{7.65}$$

7.5 2D Triangle Element

In 2D, we assume that the temperature gradient is negligible along the z-axis, $\partial T/\partial z = 0$, and nonzero in the xy-plane. The potential thus reduces to:

$$\Pi = \frac{k}{2}\int_\Omega \left(\frac{\partial T}{\partial x}\right)^2 + \left(\frac{\partial T}{\partial y}\right)^2 dV - \int_\Omega QT dV - \int_\Gamma \bar{q} T dA \tag{7.66}$$

The simplest 2D element is a triangle, which we visualize in the physical domain:

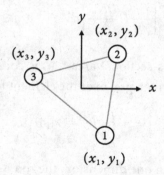

Figure 7.22. 2D triangle geometry in physical domain.

The triangular element stiffness matrix takes the following form:

$$K^e_{pq} = k\int_{\Omega^e} \left(\frac{\partial N_p}{\partial x}\frac{\partial N_q}{\partial x} + \frac{\partial N_p}{\partial y}\frac{\partial N_q}{\partial y}\right)dV; \quad p,q \in [1:3] \tag{7.67}$$

If we assume that the thickness, t, is constant, the element stiffness matrix can be reduced to an integral over the element area:

$$\left[K^e\right] = t \int_{A^e} [B]^T [D][B]\, dA \qquad (7.68)$$

To simplify calculations, we map the element from the physical to the parent domain:

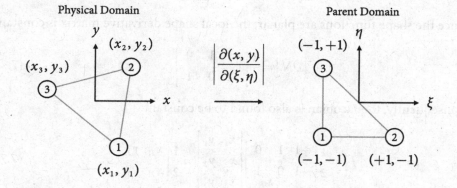

Figure 7.23. Mapping from physical to parent domain.

This mapping is reproduced in the integration via a variable change:

$$\left[K^e\right] = t \int_{\bar{A}^e} |J|[B]^T [D][B]\, d\bar{A} \qquad (7.69)$$

We recall that global coordinates can be expressed as the sum of the local shape functions multiplied by the nodal coordinates:

$$x(\xi, \eta) = \sum_{i=1:nen} x_i \bar{N}_i^e (\xi, \eta) \qquad (7.70)$$

Consequently, the directional derivative of any global coordinate is the sum of the derivatives of the local shape function multiplied by the nodal coordinates:

$$\frac{\partial x}{\partial \xi} = \sum_{i=1:nen} x_i \frac{\partial \bar{N}_i^e}{\partial \xi} = x_1 \frac{\partial \bar{N}_1^e}{\partial \xi} + x_2 \frac{\partial \bar{N}_2^e}{\partial \xi} + x_3 \frac{\partial \bar{N}_3^e}{\partial \xi} \qquad (7.71)$$

These directional derivatives provide all of the components of the Jacobian, which we express conveniently as the product of the **local shape derivative matrix**, $[\partial \bar{N}]$, and the **element coordinate matrix**, $[x]$:

$$[J] = \begin{bmatrix} \dfrac{\partial x}{\partial \xi} & \dfrac{\partial y}{\partial \xi} \\ \dfrac{\partial x}{\partial \eta} & \dfrac{\partial y}{\partial \eta} \end{bmatrix} = \begin{bmatrix} \dfrac{\partial \bar{N}_1^e}{\partial \xi} & \dfrac{\partial \bar{N}_2^e}{\partial \xi} & \dfrac{\partial \bar{N}_3^e}{\partial \xi} \\ \dfrac{\partial \bar{N}_1^e}{\partial \eta} & \dfrac{\partial \bar{N}_2^e}{\partial \eta} & \dfrac{\partial \bar{N}_3^e}{\partial \eta} \end{bmatrix} \begin{bmatrix} x_1 & y_1 \\ x_2 & y_2 \\ x_3 & y_3 \end{bmatrix} = [\partial \bar{N}][x] \tag{7.72}$$

Since the shape functions are planar, the local shape derivative matrix is constant:

$$[\partial \bar{N}] = \frac{1}{2} \begin{bmatrix} -1 & 1 & 0 \\ -1 & 0 & 1 \end{bmatrix} \tag{7.73}$$

Consequently, the Jacobian is also found to be constant:

$$[J] = \frac{1}{2} \begin{bmatrix} -1 & 1 & 0 \\ -1 & 0 & 1 \end{bmatrix} \begin{bmatrix} x_1 & y_1 \\ x_2 & y_2 \\ x_3 & y_3 \end{bmatrix} = \frac{1}{2} \begin{bmatrix} x_{21} & y_{21} \\ x_{31} & y_{31} \end{bmatrix} \tag{7.74}$$

For conciseness of presentation, we use the following shorthand notation:

$$x_{ij} = x_i - x_j; \quad y_{ij} = y_i - y_j \tag{7.75}$$

We can use this notation to explicitly express the determinant of the Jacobian:

$$|J| = \frac{x_{21} y_{31} - x_{31} y_{21}}{4} \tag{7.76}$$

It is worth noting that the determinant of the Jacobian is equal to half the area of the triangle in the physical domain, $|J| = \frac{1}{2} A$.

We next define the triangle B-matrix, which we can express as the product of the inverse of the Jacobian and the local shape derivative matrix:

$$[B] = \begin{bmatrix} \dfrac{\partial N_i^e}{\partial x} \\ \dfrac{\partial N_i^e}{\partial y} \end{bmatrix} = \begin{bmatrix} \dfrac{\partial \xi}{\partial x} & \dfrac{\partial \eta}{\partial x} \\ \dfrac{\partial \xi}{\partial y} & \dfrac{\partial \eta}{\partial y} \end{bmatrix} \begin{bmatrix} \dfrac{\partial \bar{N}_i^e}{\partial \xi} \\ \dfrac{\partial \bar{N}_i^e}{\partial \eta} \end{bmatrix} = \begin{bmatrix} \dfrac{\partial x}{\partial \xi} & \dfrac{\partial y}{\partial \xi} \\ \dfrac{\partial x}{\partial \eta} & \dfrac{\partial y}{\partial \eta} \end{bmatrix}^{-1} \begin{bmatrix} \dfrac{\partial \bar{N}_i^e}{\partial \xi} \\ \dfrac{\partial \bar{N}_i^e}{\partial \eta} \end{bmatrix} = [J]^{-1}[\partial \bar{N}] \tag{7.77}$$

Like the other ingredients in the stiffness expression, the B-matrix is constant:

$$[B] = [J]^{-1}[\partial \bar{N}] = \frac{1}{4|J|}\begin{bmatrix} y_{23} & y_{31} & y_{12} \\ x_{32} & x_{13} & x_{21} \end{bmatrix} \tag{7.78}$$

For completeness, we show that our constitutive matrix takes a trivial form:

$$[D] = \begin{bmatrix} k & 0 \\ 0 & k \end{bmatrix} \tag{7.79}$$

So far, all of the components inside the integral are constants; all that remains is the area integral of a triangle with known height ($h = 2$) and width ($w = 2$):

$$\int_{\bar{A}^e} d\bar{A} = \bar{A}^e = \frac{hw}{2} = \frac{(2)(2)}{2} = 2 \tag{7.80}$$

The element stiffness matrix can thus be expressed explicitly:

$$[K^e] = 2t|J|[B]^T[D][B] \tag{7.81}$$

Plugging in our components, we arrive at a 3×3 stiffness matrix:

$$[K^e] = \frac{t}{8|J|}\begin{bmatrix} y_{23} & x_{32} \\ y_{31} & x_{13} \\ y_{12} & x_{21} \end{bmatrix}\begin{bmatrix} k & 0 \\ 0 & k \end{bmatrix}\begin{bmatrix} y_{23} & y_{31} & y_{12} \\ x_{32} & x_{13} & x_{21} \end{bmatrix} = \begin{bmatrix} K^e_{11} & K^e_{12} & K^e_{13} \\ K^e_{21} & K^e_{22} & K^e_{23} \\ K^e_{31} & K^e_{32} & K^e_{33} \end{bmatrix} \tag{7.82}$$

This matrix will be populated by six unique values:

$$\begin{Bmatrix} K^e_{11} \\ K^e_{12} \\ K^e_{13} \\ K^e_{22} \\ K^e_{23} \\ K^e_{33} \end{Bmatrix} = \frac{kt}{8|J|}\begin{Bmatrix} x^2_{32} + y^2_{23} \\ x_{32}x_{13} + y_{23}y_{31} \\ x_{32}x_{21} + y_{23}y_{12} \\ x^2_{13} + y^2_{31} \\ x_{13}x_{21} + y_{31}y_{12} \\ x^2_{21} + y^2_{12} \end{Bmatrix} \tag{7.83}$$

The remaining entries are found using symmetry ($K^e_{pq} = K^e_{qp}$).

The triangle is the first element we have investigated in MSA or FEM with more than two nodes. As a result, there are more opportunities to create element geometries that do not perform well. In general, we want to use triangles that are close to regular. Triangles that are too skewed will perform poorly and should be broken up into two well-proportioned triangles. A triangle whose nodes lie along a line is degenerated and will thus lead to a Jacobian determinant of zero.

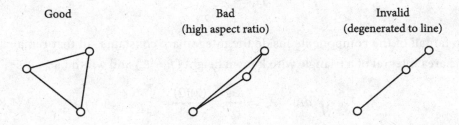

Figure 7.24.   Good, bad, and invalid triangle element geometry.

Though not necessary in our formulation, it is good practice to label nodes in counterclockwise order. Since some implementations of the triangle element use the triangle area (an absolute number) rather than the determinant of the Jacobian, labelling nodes in the opposite order will produce a negative stiffness.

7.6 Numerical Integration

Up to this point, we have examined elements whose shape functions are either linear or planar, leading to constant B-matrices and Jacobians. As a result, we were able to compute our element stiffnesses without any actual integration. Our quad, however, uses more complex shape functions defined by stretched hyperbolic paraboloid shape functions. Neither the B-matrix nor the Jacobian for our quad will be constant, and thus we will need to perform nontrivial integration.

Since we are working with approximations, we do not need to perform exact, analytical integration. Instead, we employ **numerical integration** to transform a definite integral into a weighted sum of the function at evaluation points:

$$\int_{-1}^{+1} f(\xi)d\xi = \sum_{j=1:nip} f(\xi_j)w_j \tag{7.84}$$

Gaussian Quadrature defines a set of **integration points**, ξ_j, and corresponding **integration weights**, w_j. For a 1D integral, the first three sets of integration parameters are provided in the following figure:

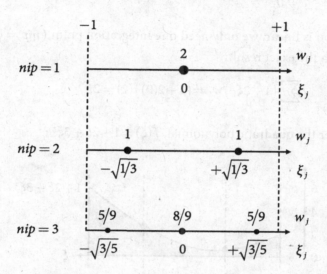

Figure 7.25. First three sets of weights and points for 1D Gaussian Quadrature.

These Gaussian Quadrature rules are tailored to polynomials. In fact, Gaussian Quadrature is guaranteed to reproduce the exact solution when we use the same number of integrations points as the highest order polynomial inside the definite integral; a linear equation is exactly integrated with one point, a quadratic with two points, etc. Gaussian Quadrature requires that the definite integral is performed within the region $\xi \in [-1 : +1]$. Fortunately, we have defined all of our local element geometries to span this range. To demonstrate 1D quadrature, consider the linear polynomial, $f(\xi) = 1 + 2\xi$:

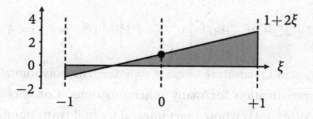

Figure 7.26. Linear polynomial evaluated at one quadrature point.

The exact definite integral is found trivially:

$$\int_{-1}^{+1} \left(1+2\xi\right) d\xi = \xi + \xi^2 \Big]_{-1}^{+1} = 2 \tag{7.85}$$

Since our function is linear, we only need one integration point ($nip = 1$, $\xi_1 = 0$, $w_1 = 2$) to match the exact result:

$$\sum_{j=1:1} \left(1+2\xi_1\right) w_1 = \left(1+2(0)\right)(2) = 2 \tag{7.86}$$

Next, consider the quadratic polynomial, $f(\xi) = 1 + 2\xi + 3\xi^2$:

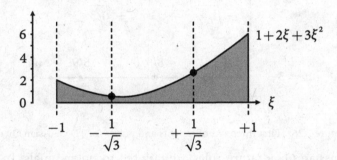

Figure 7.27. Quadratic polynomial evaluated at two quadrature points.

The exact definite integral is still relatively easy to evaluate:

$$\int_{-1}^{+1} \left(1+2\xi+3\xi^2\right) d\xi = \xi + \xi^2 + \xi^3 \Big]_{-1}^{+1} = 4 \tag{7.87}$$

Because our function is a second degree polynomial, we need to use two integration points ($nip = 2$, $\xi_j = \pm 1/\sqrt{3}$; $w_j = 1$) to match the exact solution:

$$\sum_{j=1:2} \left(1+2\xi_j+3\xi_j^2\right) w_j = \left(1-\frac{2}{\sqrt{3}}+1\right) + \left(1+\frac{2}{\sqrt{3}}+1\right) = 4 \tag{7.88}$$

Though Gaussian Quadrature is exact only for true polynomials, it can provide a good approximation for many other functions. Consider the function, $f(\xi) = \xi^2(\xi^2+5)/(\xi^2+4)$, whose exact integral we find (with significant effort):

$$\int\limits_{-1}^{+1} \left(\xi^2 \, \frac{\xi^2 + 5}{\xi^2 + 4} \right) d\xi = 0.8121 \tag{7.89}$$

Though our example function is not a polynomial, we classify it as approximately quadratic since the numerator ($\xi^4 + 5\xi^2$) is two orders higher than the denominator $\xi^2 + 4$. Thus, we can get a good approximation of the exact integration using 2-point Gaussian Quadrature:

$$\sum_{j=1:2} \left(\xi_j^2 \, \frac{\xi_j^2 + 5}{\xi_j^2 + 4} \right) w_j = 0.8205 \tag{7.90}$$

In order to use Gaussian Quadrature to integrate the quad stiffness matrix, we need to extend our integration rules to two dimensions. If we recall that an integral in 2D can be expressed as a nested integral in each dimension, all we need to do is to perform the weighted summation for each dimension:

$$\int\limits_{-1}^{+1} \int\limits_{-1}^{+1} f(\xi, \eta) d\xi d\eta = \sum_a \sum_b f(\xi_a, \eta_b) w_a w_b = \sum_{j=1:nip} f(\xi_j, \eta_j) w_j \tag{7.91}$$

Note that we reduce this expression into a single summation with one cumulative weight ($w_j = w_a w_b$). The resulting integration point distribution and weights can be visualized as follows:

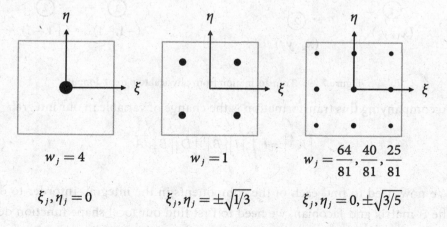

Figure 7.28. Weights and integration points for 2D Gaussian Quadrature.

7.7 2D Quad Element

As for the triangle, we assume that the temperature gradient within the quad is nonzero in the xy-plane and negligible along the z-axis, $\partial T/\partial z = 0$. The components of the stiffness equation iterate through four, rather than three nodal dofs:

$$K^e_{pq} = k \int\limits_{\Omega^e} \left(\frac{\partial N_p}{\partial x} \frac{\partial N_q}{\partial x} + \frac{\partial N_p}{\partial y} \frac{\partial N_q}{\partial y} \right) dV ; \quad p,q \in [1:4] \tag{7.92}$$

In matrix form, the element stiffness again reduces to an integral over the element area if we assume that the thickness, t, is constant:

$$\left[K^e \right] = t \int\limits_{A^e} \left[B \right]^T \left[D \right] \left[B \right] dA \tag{7.93}$$

As with the triangle, we will transform the integration to the parent domain:

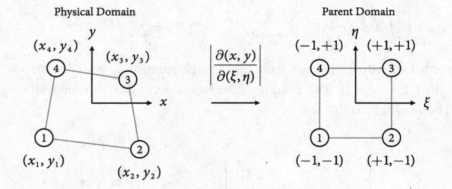

Figure 7.29. Transformation from physical to parent domain.

Accompanying this transformation is the change of variable in our integral:

$$\left[K^e \right] = t \int\limits_{\bar{A}^e} |J| \left[B \right]^T \left[D \right] \left[B \right] d\bar{A} \tag{7.94}$$

We now need to find each of the components in the integral. In order to derive the B-matrix and Jacobian, we need to first find our local shape function derivative matrix, $[\partial \bar{N}]$, which is not constant, but linear:

$$[\partial \bar{N}] = \begin{bmatrix} \dfrac{\partial \bar{N}_1^e}{\partial \xi} & \dfrac{\partial \bar{N}_2^e}{\partial \xi} & \dfrac{\partial \bar{N}_3^e}{\partial \xi} & \dfrac{\partial \bar{N}_4^e}{\partial \xi} \\ \dfrac{\partial \bar{N}_1^e}{\partial \eta} & \dfrac{\partial \bar{N}_2^e}{\partial \eta} & \dfrac{\partial \bar{N}_3^e}{\partial \eta} & \dfrac{\partial \bar{N}_4^e}{\partial \eta} \end{bmatrix} = \frac{1}{2} \begin{bmatrix} -1+\eta & 1-\eta & 1+\eta & -1-\eta \\ -1+\xi & -1-\xi & 1+\xi & 1-\xi \end{bmatrix} \quad (7.95)$$

Our element coordinate matrix is defined simply:

$$[x^e] = \begin{bmatrix} x_1 & y_1 \\ x_2 & y_2 \\ x_3 & y_3 \\ x_4 & y_4 \end{bmatrix} \quad (7.96)$$

The Jacobian and B-matrix can be calculated using these two matrices:

$$[J] = [\partial \bar{N}][x^e]; \quad [B] = [J]^{-1}[\partial \bar{N}] \quad (7.97)$$

Lastly, we recall that the constitutive matrix is still trivially defined:

$$[D] = \begin{bmatrix} k & 0 \\ 0 & k \end{bmatrix} \quad (7.98)$$

Having found expressions for all of the components inside the element stiffness expression, we next undertake the integration. Since we are working in the parent domain, our definite integral conveniently spans $\xi, \eta \in [-1 : +1]$:

$$[K^e] = t \int_{-1}^{+1} \int_{-1}^{+1} |J(\xi, \eta)| [B(\xi, \eta)]^T [D] [B(\xi, \eta)] d\xi d\eta \quad (7.99)$$

Other than the constitutive matrix, the components inside the integral are not constants and thus cannot be removed from the integral. Using the numerical integration rules from the preceding section, we transform the continuous integral into a weighted sum:

$$[K^e] = t \sum_{j=1:nip} |J(\xi_j, \eta_j)| [B(\xi_j, \eta_j)]^T [D] [B(\xi_j, \eta_j)] w_j \quad (7.100)$$

The expression inside the integral is not a true polynomial; in order to choose an appropriate number of integration points, we need to establish the approximate degree of the integrand. We observe that the determinant of the Jacobian is quadratic, while the B-matrix includes a quadratic polynomial divided by a quadratic term; the product of the Jacobian and two B-matrices is thus approximately quadratic. Because this integrand may be quadratic in either dimension, we elect to use 2×2 Gaussian Quadrature (four points):

$$nip = 2 \times 2 = 4; \quad \left(\xi_j, \eta_j \right) = \left(\pm \frac{1}{\sqrt{3}}, \pm \frac{1}{\sqrt{3}} \right); \quad w_j = 1 \tag{7.101}$$

Numerical integration with four points becomes exact for rectangular quads; the transformation between the parent and physical domains is linear in both directions, thus generating a constant Jacobian.

As with the triangle, we want to use well-proportioned, convex elements. Because we now have four nodes defining the quad geometry, we need to be even more diligent to avoid bad or invalid geometries:

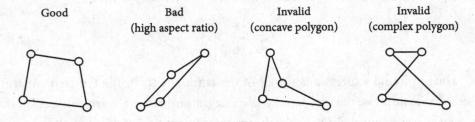

Figure 7.30.   Good, bad, and invalid quad element geometries.

7.8   Applying Boundary Conditions

The elements developed in this chapter can be assembled into a global matrix using the assembly techniques we established in MSA. In order to perform an analysis, we still need to define the displacement and force vectors acting on the global mesh. The essential BCs must be turned into prescribed displacement or temperature vectors, while the natural BCs must be turned into applied force or flux vectors. In this section, we will develop a systematic approach for transforming both types of continuous BC into discretized vectors.

7.8.1 Essential Boundary Conditions

In the original BVP, the essential BCs take the form of a continuous displacement or temperature profile along a region of the bounding surface. The most reliable way to discretize this profile is to set nodal displacements/temperatures to the value of the continuous BC at each node:

$$\tilde{T}_s(x_n, y_n) = T_s(x_n, y_n) \tag{7.102}$$

This approach leads to an approximation of the continuous temperature profile which improves with the resolution of the mesh.

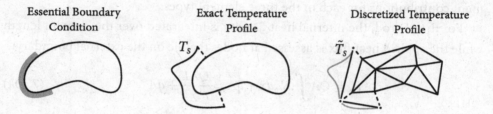

| Essential Boundary Condition | Exact Temperature Profile | Discretized Temperature Profile |

Figure 7.31. Approximating an essential BC as a discretized temperature profile.

7.8.2 Natural Boundary Conditions

The applied continuous natural BCs must be similarly converted into discretized nodal values, as visualized below.

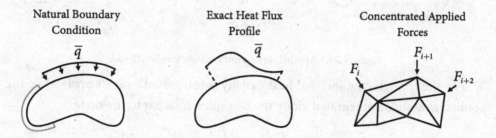

| Natural Boundary Condition | Exact Heat Flux Profile | Concentrated Applied Forces |

Figure 7.32. Approximating a continuous natural BC as concentrated applied forces/fluxes.

Natural BCs typically take the form of distributed forces or fluxes, which are transformed into concentrated nodal forces through integration, typically on an element-by-element basis. We recall that the force contribution at each element node is the sum of a domain and a boundary integral:

$$F_p^e = \int_{\Omega^e} Q N_p^e dV + \int_{\Gamma^e} \overline{q} N_p^e dA \qquad (7.103)$$

The integral over the element boundary only pertains to elements adjacent to the BVP boundary; an internal element will experience only internal fluxes. We assume that both internal and applied heat sources are constant and thus can be moved out of the integral; finer meshing will converge to capture any non-uniformity. Using this assumption, we derive the concentrated applied force element contributions for each of the three element types.

For the 1D rod, the internal heat density is integrated over the element length while the applied heat flux is assessed at nodes that lie on the external boundary:

$$F_p^e = QA \int_{L^e} N_p^e dx + \overline{q}A = \frac{QAL}{2} + \overline{q}A \qquad (7.104)$$

We note that the total internal heat energy generated, QAL, is split equally between the two nodes. The applied heat flux for a rod can only be applied at rod ends and must be integrated over the rod cross-sectional area:

Element Geometry	Internal Heat Source Contribution	External Heat Flux Contribution

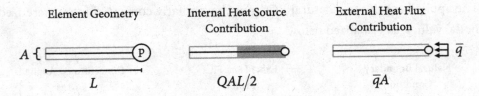

L	$QAL/2$	$\overline{q}A$

Figure 7.33. Applied force contributions for the 1D rod.

For the triangle, the internal heat density is integrated over its area, while the applied heat flux is integrated along the two sides adjacent to the node:

$$F_p^e = Qt \int_{A^e} N_p^e dA + \overline{q}t \int_{s^e} N_p^e ds = \frac{QAt}{3} + \frac{\overline{q}_1 L_1 t}{2} + \frac{\overline{q}_2 L_2 t}{2} \qquad (7.105)$$

The internal heat, QAt, is distributed equally between all three nodes. The applied heat flux is distributed equally to both nodes of any side:

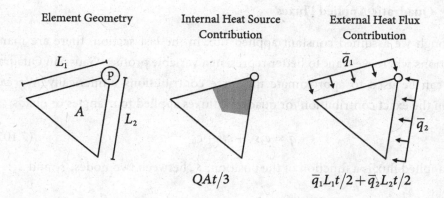

Figure 7.34. Applied force contributions for 2D triangle.

The force contribution for the quad closely parallels that of the triangle:

$$F_p^e = Qt \int_{A^e} N_p^e dA + \overline{q}t \int_{s^e} N_p^e ds \approx \frac{QAt}{4} + \frac{\overline{q}_1 L_1 t}{2} + \frac{\overline{q}_2 L_2 t}{2} \qquad (7.106)$$

For an irregular quad, the integral of the heat density is actually quite difficult to perform. We could use 2×2 Gaussian Quadrature to find an approximate integral, but we can also assume that the total internal heat is distributed equally to each of the four nodes. This approximation is very good for well-proportioned quads and exact for rectangles. The applied heat flux applied to any side is equally distributed to each node on that side:

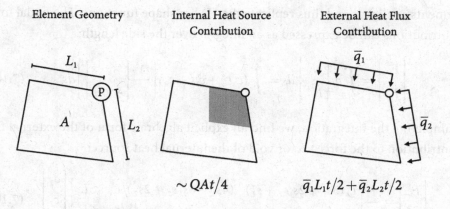

Figure 7.35. Applied force contributions for 2D quad.

7.8.3 Quadratic Applied Fluxes

Although we assumed constant applied flux in the last section, there are many situations where we want to better represent a variable profile. Gaussian Quadrature can be used to approximate the force contributions numerically. We can obtain the exact contribution for quadratic fluxes applied to triangles or quads:

$$\bar{q} = c_2 s^2 + c_1 s + c_0 \tag{7.107}$$

The applied flux is a function of the position, s, between two nodes, s_1 and s_2:

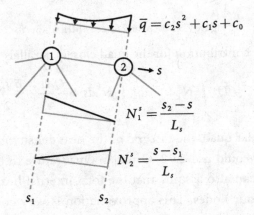

Figure 7.36. Applied force contributions for 2D quad.

The two edge **perimeter shape functions**, N_1^s and N_2^s, for both triangle or quad elements are linear and thus replicate the 1D bar shape functions. The nodal force contributions can be expressed as an integral over the side length:

$$\begin{Bmatrix} F_1^e \\ F_2^e \end{Bmatrix} = \int_{L_s} \bar{q} \begin{Bmatrix} N_1^s \\ N_2^s \end{Bmatrix} ds = \int_{s_1}^{s_2} (c_2 s^2 + c_1 s + c_0) \left(\frac{1}{L_s} \begin{Bmatrix} s_2 - s \\ s - s_1 \end{Bmatrix} \right) ds \tag{7.108}$$

Completing the integration, we find an explicit algebraic form of the external flux contribution to the force vector void of the internal heat source:

$$\begin{Bmatrix} F_1^e \\ F_2^e \end{Bmatrix} = \frac{t}{12} \begin{bmatrix} L_s(3s_1^2 + 2s_1 s_2 + s_2^2) & (4s_1^3 - 6s_1^2 s_2 + 2s_2^3)/L_s & 6L_s \\ L_s(s_1^2 + 2s_1 s_2 + 3s_2^2) & (2s_1^3 - 6s_1 s_2^2 + 4s_2^3)/L_s & 6L_s \end{bmatrix} \begin{Bmatrix} c_2 \\ c_1 \\ c_0 \end{Bmatrix} \tag{7.109}$$

7.9 Code Modifications

Though the derivation of FEM elements is quite a bit more intricate than MSA elements, their implementation into code follows nearly the same approach. In addition to making the expected changes to our element definition function and writing a new element function, we also need to introduce two new subroutines for calculating shape functions and generating Gaussian Quadrature points, which will prove useful in our elasticity and plate/shell chapters. We also introduce functions for mapping between physical and parent domains and integrating a distributed parabolic flux applied to the side of a triangle or quad element.

7.9.1 Generate Shape Functions

The genShape function is the first of two functions unique to FEM. This function provides several components necessary to the calculation of our element stiffness matrix (B-matrix, B, and the determinant of the Jacobian, detJ) as well as the complete set of shape functions, N, and the local shape derivative matrix, dN. This function is provided with two inputs: the element coordinates, xe, and the evaluation point in the parent domain, xi (because ξ is pronounced "xi").

```
1 function [B,detJ,N,dN] = genShape(xe,xi)
2
3 nen = size(xe,1);                              % number of element nodes
4 switch nen
5   case 2
6     N  = [1-xi(1) 1+xi(1)]/2;                            % shape function, N
7     dN = [-1 1]/2;        % partial derivatives of shape functions, dN
8   case 3
9     N  = [-xi(1)-xi(2) 1+xi(1) 1+xi(2)]/2;
10    dN = [-1 1 0; -1 0 1]/2;
11  case 4
12    N  = (1+xi(1)*[-1 1 1 -1]).*(1+xi(2)*[-1 -1 1 1]);
13    dN = ([-1 1 1 -1;-1 -1 1 1]+xi'*[1 -1 1 -1])/4;
14 end
15
16 J = dN*xe;                                    % calculate jacobian, J
17 B = J\dN;                              % calculate general B-matrix, B
18 detJ = det(J);            % calculate determinant of jacobian, detJ
```

We will use this function to generate the B-matrix and the Jacobian determinant for our stiffness matrix calculation. For the rod and triangle these components are constants, so an arbitrary point for xi may be supplied.

As evident from the structure of this function, the type of element is distinguished by the number of element nodes. It may not be possible to distinguish more complex, higher-dimensional elements just using this parameter and it will be necessary to use a specific input field to denote elements. For instance, the simplest 3D element is a tetrahedron with the same number of nodes (four) as the quad. Additionally, there is an entire class of elements with nodes between vertices, further necessitating a precise distinction between elements.

7.9.2 Generate Gauss Integration Points

The **genGauss** function is the second function unique to FEM. This function generates the integration points (pts) and weights (wts) for evaluating a 2D integral in the parent domain. This function is supplied the total number of integration points (nip) corresponding to 1×1 (nip = 1), 2×2 (nip = 4), or 3×3 (nip = 9) integration schemes.

```
 1 function [pts,wts] = genGauss(nip)
 2
 3 switch nip
 4    case 1
 5       wts = 4;
 6       pts = [0 0];
 7    case 4
 8       wts = [1 1 1 1];
 9       pts = [-1 -1;1 -1;1 1;-1 1]*sqrt(1/3);
10    case 9
11       wts = [25 40 25 40 64 40 25 40 25]/81;
12       pts = [-1 -1;0 -1;1 -1;-1 0;0 0;1 0;-1 1;0 1;1 1]*sqrt(3/5);
13 end
```

As with **genGauss**, this function can be updated to work with other dimensions, more points, or nonhomogeneous (i.e., 1×2) integration schemes.

7.9.3 Modifications to the Element Definition Function

The `defElems` function is altered by replicating the familiar six lines and updating the element function name (`Ke_heat`) and the active dofs in each dimension:

```
% 4. Heat
net = net + 1;
kList{typ} = 'Ke_heat';
iad(net,:,1) = [0 0 0 0 0 0 1]; % 1D
iad(net,:,2) = [0 0 0 0 0 0 1]; % 2D
iad(net,:,3) = [0 0 0 0 0 0 0]; % 3D
```

7.9.4 Heat Element

Using the two new functions, the heat element, `Ke_heat`, is quite easy to generate. Once again, the number of element nodes differentiates between element types.

```
 1 function [Ke,D,Te] = Ke_heat(xe,prop)
 2
 3 nen = size(xe,1);                    % number of element nodes
 4 nsd = size(xe,2);                    % number of spatial dimensions
 5 nph = prop(11);                      % number of heat integration points
 6
 7 D = prop(8)*eye(nsd);
 8 ke = zeros(nen);
 9
10 switch nen
11   case 2                             % if rod
12     [B,detJ] = genShape(xe,0);
13     ke = prop(2)*detJ*B'*D*B;
14   case 3                             % if triangle
15     [B,detJ] = genShape(xe,[0 0]);
16     ke = prop(7)*detJ*B'*D*B*2;
17   case 4                             % if quad
18     [pts,wts] = genGauss(nph);
19     for i = 1:size(pts,1)
20       [B,detJ] = genShape(xe,pts(i,:));
21       ke = ke + prop(7)*detJ*B'*D*B*wts(i);
22     end
23 end
24
25 Te = 1;
26 Ke = Te'*ke*Te;
```

There are several parts of this function worth highlighting. The stiffness expression for the rod uses the element area (stored in prop(2)), while the triangle and quad use the element thickness (stored in prop(7)). The rod and triangle stiffnesses are found explicitly while the quad element includes a loop to assemble the stiffness matrix using contributions from each integration point. The element constitutive matrix, D, supplied in the output field, is typically reserved for the local element stiffness matrix, ke. For consistency, the rod stiffness is obtained using integration in the parent domain:

$$\left[K^e \right] = A \left| J \right| \left[B \right]^T \left[D \right] \left[B \right] \tag{7.110}$$

The local shape function derivative matrix and the coordinate vector are simple:

$$\left[\partial \bar{N} \right] = \frac{1}{2} \left[-1 \; +1 \right]; \quad \left[x \right] = \begin{bmatrix} x_1 \\ x_2 \end{bmatrix} \tag{7.111}$$

The Jacobian (and its determinant in 1D) is equal to half the rod length:

$$\left[J \right] = \left[\partial \bar{N} \right] \left[x \right] = \frac{1}{2} \left(x_2 - x_1 \right) \tag{7.112}$$

7.9.5 Integrating the Force Vector

We introduce the function, **intForce**, to integrate a quadratic flux profile. This function is supplied with two coordinates along the side length (s), a vector containing the coefficients in the quadratic flux profile (c), element thickness (t), and an optional parameter (node), designating if a force at one node is to be returned.

```
 1 function Fe = intForce(s,c,t,node)
 2
 3 s1 = s(1); s2 = s(2);                                    % side ordinates
 4 Ls = abs(s2-s1);                                         % side length
 5
 6 Fe = [Ls*(3*s1^2+2*s1*s2+s2^2) (4*s1^3-6*s1^2*s2+2*s2^3)/Ls 6*Ls;
 7      Ls*(s1^2+2*s1*s2+3*s2^2) (2*s1^3-6*s1*s2^2+4*s2^3)/Ls 6*Ls];
 8 Fe = t/12*Fe*c';
 9
10 % Output only one force if node selection provided
11 if exist('node'), Fe = Fe(node); end
```

7.9.6 Mapping between Domains

Though not immediately necessary, we also want to establish a function for forward and backward mapping between parent and physical domains. Mapping from the parent to the physical domain is trivially achieved in two lines, so we do not write a separate function:

```
[~,~,N] = genShape(xe,xi)
x = N*xe;
```

Since backward mapping is a bit more complicated, we need to write a function, mapParent, which implements the mapping equations we derived earlier in the chapter. The number of nodes is once again used to distinguish between elements.

```
 1 function [xi] = mapParent(xe,x)
 2
 3 switch size(xe,1)                              % number of element nodes
 4   case 2
 5     xi = (2*x-xe(1)-xe(2))/(xe(2)-xe(1));
 6   case 3
 7     x1 = xe(1,1); x2 = xe(2,1); x3 = xe(3,1);
 8     y1 = xe(1,2); y2 = xe(2,2); y3 = xe(3,2);
 9     c2 = x1*(y2-y3)+x2*(y3-y1)+x3*(y1-y2);
10
11     xi(1) =((y3-y1)*(2*x(1)-x2)+(x1-x3)*(2*x(2)-y2)-x1*y3+x3*y1)/c2;
12     xi(2) =((y1-y2)*(2*x(1)-x3)+(x2-x1)*(2*x(2)-y3)+x1*y2-x2*y1)/c2;
13   case 4
14     a = [1 1 1 1;-1 1 1 -1;-1 -1 1 1;1 -1 1 -1]*xe(:,1);
15     b = [1 1 1 1;-1 1 1 -1;-1 -1 1 1;1 -1 1 -1]*xe(:,2);
16
17     c1 = a(2)*b(4)-a(4)*b(2);
18     c2 = a(4)*b(3)-a(3)*b(4);
19     c3 = a(2)*b(3)-a(3)*b(2);
20
21     f1 =-a(4)*(b(1)-4*x(2))+b(4)*(a(1)-4*x(1))+c3;
22     f2 =-a(3)*(b(1)-4*x(2))+b(3)*(a(1)-4*x(1));
23     f3 = a(4)*(b(1)-4*x(2))-b(4)*(a(1)-4*x(1))+c3;
24     f4 = a(2)*(b(1)-4*x(2))-b(2)*(a(1)-4*x(1));
25
26     if c1 ~= 0, xi(1) =(-f1 + sqrt(f1^2 - 4*c1*f2))/(2*c1);
27     else        xi(1) = -f2/f1; end
28     if c2 ~= 0, xi(2) =(-f3 + sqrt(f3^2 - 4*c2*f4))/(2*c2);
29     else        xi(2) = -f4/f3; end
30 end
```

7.10 Example

To demonstrate how to solve a heat BVP, we consider the following rectangular problem with no internal heat source ($Q = 0$):

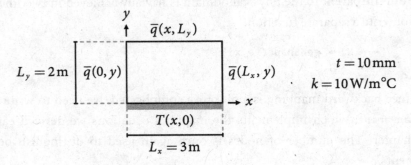

Figure 7.37. Heat example set up.

Three sides experience trivial natural BCs in the form of zero heat fluxes:

$$\bar{q}(0, y) = \bar{q}(x, L_y) = \bar{q}(L_x, y) = 0 \tag{7.113}$$

Along the bottom of the rectangle, we prescribe an essential BC in the form of a sinusoidal temperature distribution:

$$T(x, 0) = T_0 \left(1 + \cos\left(\pi x / L_x \right) \right); \quad T_0 = 10°C \tag{7.114}$$

For this specific example, the exact solution is supplied by Carslaw and Jaeger:

$$T(x, 0) = T_0 \left[1 + \frac{\cos\left(\pi x / L_x \right) \cosh\left(\pi (L_y - y)/L_x \right)}{\cosh\left(\pi L_y / L_x \right)} \right] \tag{7.115}$$

In the exact solution, the temperature reaches its maximum ($20°C$) at the bottom left corner and its minimum ($0°C$) at the bottom right corner:

0°C ▭ 20°C

Figure 7.38. Exact temperature profile.

7.10.1 Meshing Algorithm

When we transform a continuous BVP into a discretized FEM problem, we have countless meshing options available to us. If we are consistent in using well-formulated, geometrically-regular elements, we should observe an improvement in the approximation as we increase the number of elements. As a result, we typically use a meshing strategy rather than a single mesh; for this problem, we employ a rectangular **structured** mesh. The `genMesh` function generates a mesh defined by nodal coordinates, xn, and the element connectivity matrix, ien, for a rectangle dimensioned Lx by Ly.

```
1 function [xn,ien] = genMesh(Lx,Ly,neX,neY,nsd,mesh)
2
3 nel = neX*neY;                               % number of elements
4 if mesh < 3, nen = 3; nel = nel*2;
5 else          nen = 4; end
6 nnp = (neX+1)*(neY+1);                       % number of nodal points
7
8 % nodal definitions
9 Xinc = Lx/neX;  Yinc = Ly/neY;
10 xn = zeros(nnp,nsd);
11 for i = 1:neX+1
12   for j = 1:neY+1
13     n = i+(j-1)*(neX+1);
14     xn(n,:) = [(i-1)*Xinc (j-1)*Yinc];
15   end
16 end
17
18 % element definitions
19 ien = zeros(nel,nen);                        % index of element nodes
20 for i = 1:neX
21   for j = 1:neY
22     e = i+(j-1)*neX;
23     n1 = i+(j-1)*(neX+1); n2 = n1 + 1;
24     n3 = i+j*(neX+1); n4 = n3 + 1;
25     switch mesh
26       case 1, ien(2*e-1,:) = [n1 n2 n4];
27               ien(2*e,:)   = [n1 n4 n3];
28       case 2, ien(2*e-1,:) = [n1 n2 n3];
29               ien(2*e,:)   = [n2 n4 n3];
30       case 3, ien(e,:)     = [n1 n2 n4 n3];
31     end
32   end
33 end
```

The mesh input controls the type of mesh (right biased triangle, left biased triangle, or quad):

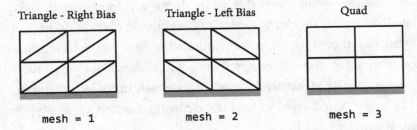

Figure 7.39. Mesh types.

The number of elements per side (neX and neY) controls the number of elements or mesh **refinement**:

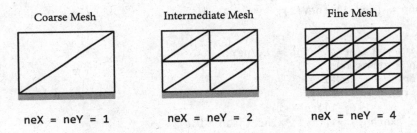

Figure 7.40. Mesh refinement.

The meshing algorithm marches through the horizontal (i) and vertical (j) nodes to find the four perimeter nodes (n1, n2, n3, and n4) of a quad:

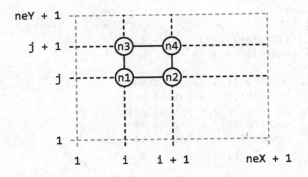

Figure 7.41. Nodal numbering.

Based on the type of mesh employed, the rectangle is either subdivided into two triangles or directly implemented as a quad using the following element ordering strategies:

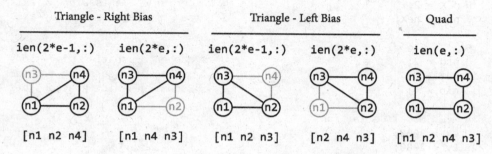

Figure 7.42. Element connectivity numbering.

To demonstrate this meshing procedure, we diagram the final mesh for a 2×2 right-biased triangle:

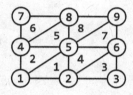

Figure 7.43. Example nodal and element numbering for `mesh = 1` and `neX = neY = 2`.

A regular mesh is not always the best choice for an analysis as it frequently produces skewed results. Irregular meshes, such as those generated using **Delaunay triangulation**, are best suited for avoiding bias in the analysis.

7.10.2 Main Script

Using our meshing algorithm, we write a simple script to execute our analysis. We need to define a new set of inputs (`mesh`, `neX`, and `neY`) which we will modify directly in the main script for our convergence studies. After generating the mesh, we apply our essential BCs by iterating through all of the nodes, checking if the node lies on the x-axis, and then assigning to it a prescribed temperature using the provided temperature profile.

```
1 % exChpt7 - Analysis of 2D heat problem in degC and mm base units
2 clear;
3
4 % 0. Convergence inputs
5 mesh = 2;              % mesh type, 1 = right tri, 2 = left tri, 3 = quad
6 neX = 2;                             % number of elements along x-axis
7 neY = neX;                           % number of elements along y-axis
8
9 % 1. Global definitions
10 nsd = 2;                            % number of spatial dimensions
11 ndf = 7;                            % number of degrees of freedom
12 nip = 4;                            % number of integrations pts
13
14 % 2. Nodal definitions
15 Lx = 3000;
16 Ly = 2000;
17 T0 = 10;
18 k = 0.01;                           % thermal conductivity, kh
19 t = 10;                                   % thickness, t
20
21 [xn,ien] = genMesh(Lx,Ly,neX,neY,nsd,mesh);
22 nnp = size(xn,1);
23 nel = size(ien,1);
24 idb = zeros(nnp,ndf);                   % index of dofs - supported
25 ds = zeros(nnp,ndf);       % prescribed displacements at supports (C)
26 Pu = zeros(nnp,ndf);                        % applied forces (f)
27
28 for n = 1:nnp
29   if xn(n,2) == 0
30     idb(n,:) = 1;
31     ds(n,7) = T0*(1+cos(pi*xn(n,1)/Lx))
32   end
33 end
34
35 prop = repmat([4 0 0 0 0 0 0 t k 0 0 nip 0 0 0 0],[nel 1]);
36
37 % 4. RUN ANALYSIS
38 [results,process] = runAnalysis(Pu,ds,xn,prop,idb,ien);
39 [F,Rs,Fe,Fi,d,du,de] = deal(results{:});
40 [Kuu,Ke,ke,Te,ied,idu,ids] = deal(process{:});
```

7.10.3 Results

Interpreting large amounts of data poses a difficulty inherent to FEM analysis. Even though our example problem is simple by FEM standards, the various meshes quickly accumulate an incomprehensible data set. In this section, we demonstrate several techniques for clearly displaying FEM results. We will also use the exact solution to evaluate the aptitude of the approximations through convergence and error studies.

7.10.3.1 Temperature

Because the number of elements in this example is small, we are able to present all of the nodal temperatures. The nodal temperatures in the two triangular meshes are incidentally identical so we present only one set of results for both.

	1×1 Mesh			2×2 Mesh			4×4 Mesh		Exact Sol'n
	Tri	Quad		Tri	Quad		Tri	Quad	
n	T(°C)	T(°C)	n	T(°C)	T(°C)	n	T(°C)	T(°C)	T(°C)
1	20.00	20.00	1	20.00	20.00	1	20.00	20.00	20.00
						2	17.07	17.07	17.07
			2	10.00	10.00	3	10.00	10.00	10.00
						4	2.93	2.93	2.93
2	0.00	0.00	3	0.00	0.00	5	0.00	0.00	0.00
						6	16.22	15.95	16.09
						7	14.40	14.21	14.30
						8	10.00	10.00	10.00
						9	5.60	5.79	5.70
						10	3.78	4.05	3.91
			4	14.55	13.15	11	14.06	13.71	13.88
						12	12.87	12.62	12.75
			5	10.00	10.00	13	10.00	10.00	10.00
						14	7.13	7.38	7.25
			6	5.45	6.85	15	5.94	6.29	6.12
						16	12.95	12.58	12.77
						17	12.08	11.83	11.96
						18	10.00	10.00	10.00
						19	7.92	8.17	8.04
						20	7.05	7.42	7.23
3	15.29	10.40	7	13.15	11.70	21	12.61	12.24	12.43
						22	11.84	11.59	11.72
			8	10.00	10.00	23	10.00	10.00	10.00
						24	8.16	8.41	8.28
4	4.71	9.60	9	6.85	8.30	25	7.39	7.76	7.57

Even with the efficiency of our tabulation, it is difficult to make sense of the temperature distribution or how it compares to the exact solution. In general, FEM results need to be assessed both numerically and graphically. Because we are working with a continuum, we want to observe the temperature distribution over the entire domain. The following plots were generated by evaluating the temperature at the center of each pixel in a plot and then assigning it a grayscale value (white corresponding to $0°C$ and black corresponding to $20°C$). The results are ordered vertically by mesh refinement, with the coarsest mesh at the top.

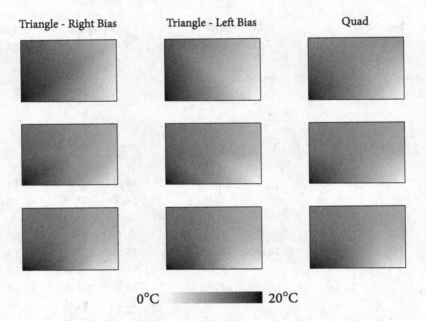

Figure 7.44. Temperature distribution for all three mesh types and refinements.

We observe immediately that the temperature distribution among all nine meshes demonstrates a consistent distribution, but it is still very difficult to assess the relative performance of each mesh. Although the two triangular meshes share nodal temperatures, their basis functions are different resulting in a noticeably different temperature distribution within the elements.

We can get a better understanding of the improvement in performance by plotting the temperature profile along the perimeter, s, of the BVP domain:

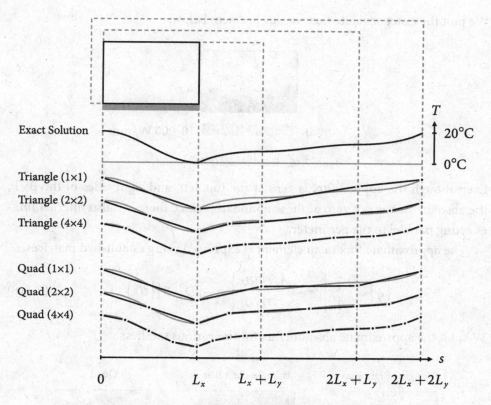

Figure 7.45. Convergence of perimeter temperatures.

This graphic demonstrates the correlation between approximation accuracy and mesh refinement for both triangle and quad meshes.

7.10.3.2 Flux

To characterize the flux within the system, we use the **absolute flux**, $|q|$:

$$|q| = \sqrt{q_x^2 + q_y^2} \tag{7.116}$$

Since we have the exact temperature available for our BVP, we can also obtain the exact directional heat fluxes, which supply the components of the absolute flux:

$$\{q\} = \begin{Bmatrix} q_x \\ q_y \end{Bmatrix} = -\frac{T_0 \pi}{L_x \cosh\left(\pi L_y / L_x\right)} \begin{Bmatrix} \sin\left(\pi x / L_x\right) \cosh\left(\pi (L_y - y)/L_x\right) \\ \cos\left(\pi x / L_x\right) \sinh\left(\pi (L_y - y)/L_x\right) \end{Bmatrix} \tag{7.117}$$

We plot the exact absolute flux for our problem below:

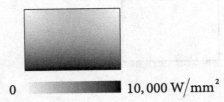

Figure 7.46. Exact absolute flux profile.

Even though the applied flux is zero at the top, left, and right sides of the BVP, the absolute flux is not zero at these boundaries since there are also internal fluxes acting parallel to the perimeter.

The approximate flux in an element is expressed using established matrices:

$$\{\tilde{q}\} = \begin{Bmatrix} \tilde{q}_x \\ \tilde{q}_y \end{Bmatrix} = -k \begin{Bmatrix} \partial \tilde{T}/\partial x \\ \partial \tilde{T}/\partial y \end{Bmatrix} = -[D][B]\{d^e\} \tag{7.118}$$

We plot the approximate absolute flux for the various meshes:

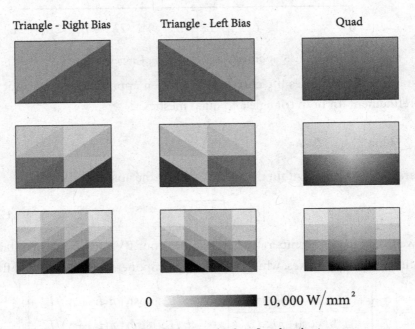

Figure 7.47. Approximate absolute flux distributions.

We note that these fluxes provide a relatively poor approximation of the exact fluxes both at nodes and element boundaries. The flux, unlike temperature, shows discontinuities across element boundaries. These discontinuities arise because the approximate temperature is only C^0 continuous; since flux is defined as a derivative of temperature, it does not exist at element boundaries. The fluxes inside the elements are either constant (triangle) or linear (quad) as defined by the element B-matrices.

The first derivative (flux or stress) of the primary function (temperature or displacement) in most FEM problems typically demonstrates these discontinuities. In the next chapter, we will examine how smoothing strategies may be used to resolve these discontinuities.

7.10.3.3 Convergence and Error

Convergence is an investigation of how approximate solutions improve with mesh resolution. A precise way of assessing the convergence of a specific problem is to track the value at one critical dof over the various meshes. For our example, the temperature at the top-right corner of our mesh is one such critical dof which we plot for the three resolutions of the triangle and quad meshes:

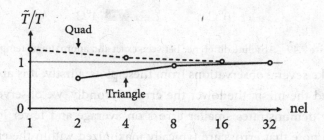

Figure 7.48. Convergence of temperature at the top-right corner ($L_x.L_y$).

For convergence, it is convenient to plot the number of elements along a logarithmic scale since the mesh increases by squares of two. For a $n \times n$ mesh, there are n^2 quad elements and twice as many ($2n^2$) triangular elements, thus the triangle and quad convergence curves are offset from each other. As expected, the quad mesh converges more quickly than the triangle mesh.

Because the exact solution is available, we also have the unique opportunity to perform an **error analysis**. While atypical, this analysis helps to demonstrate some unique characteristics of an approximation. We visualize the error in the temperature by plotting the absolute difference between the exact and approximate solutions:

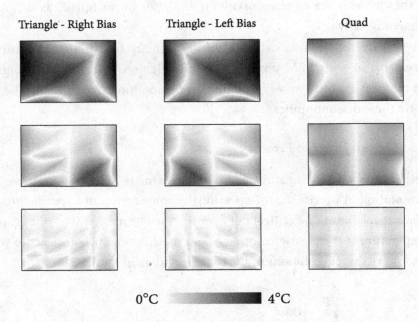

Figure 7.49.   Absolute difference between exact and approximate temperature.

We can make several observations from these plots. Firstly, it is apparent that the more refined the mesh, the lower the error. Secondly, we observe that the quad consistently demonstrates smaller errors on average and fewer localized spikes. Thirdly, we note that errors are typically maximized within the element domain rather than at the nodes.

The last observation is demonstrative of a general principle of Variational Principles, whereby the approximate solution is best at the nodes while the derivatives behave better within the element domain. Since Fourier's law defines the flux as linearly related to the temperature gradient, flux is effectively a derivative of the temperature. To demonstrate the second part of this principle, we plot the error in the absolute flux:

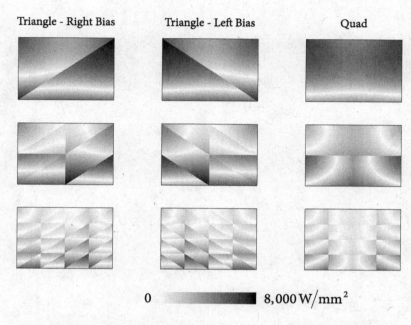

Figure 7.50. Absolute difference between exact and approximate absolute flux.

Chapter 8

Linear Elasticity

Having established a basic framework for the implementation of FEM via steady-state heat conduction, we move on to the more advanced problem of static linear elasticity. This increase in complexity is a consequence of the primary unknown taking the form of a displacement vector field, rather than a scalar temperature distribution. Linear elasticity is governed by the same three relationships (equilibrium, constitutive, and kinematic) that we used in the derivation of the truss and beam elements but articulated using three-dimensional directional derivatives.

While we will present the overarching concepts in three dimensions, we will limit new element derivations to the two-dimensional elastic triangle and quad elements. As a pedagogical aid, we also derive the 1D truss element using FEM to demonstrate equivalence between MSA and FEM. Upon establishing these elements, we also demonstrate how to recover strains and stresses. Since strains and stresses are derivatives of a primary function like heat fluxes, they will be discontinuous across element boundaries. Hence, we develop a technique for generating continuous strain, stress, or flux profiles using stress smoothing. Next, we introduce the necessary code modifications for implementing the new elements, extracting strains and stresses, and stress smoothing. We conclude with an illustrative analysis of a deep cantilever.

8.1 Strong Form of the BVP

We begin by presenting the strong form of the elastic BVP as defined by the governing equations for linear elasticity and the boundary conditions.

8.1.1 Governing Equations

The governing relationships for linear elasticity are quite complex in three dimensions. We investigate these relationships in reference to an infinitesimal cube within a continuous elastic medium:

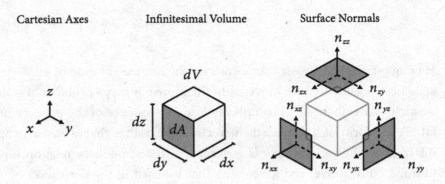

Figure 8.1. Geometry of an infinitesimal volume.

In addition to defining the cube geometry, we also specify the local axes for each surface of the cube. The components of these axes are labelled n_{ij} with indices i and j identifying the surface and axis orientations respectively; when the surface and axis orientations align ($i = j$), we obtain the surface normal, n_i. There is no distinction in notation for axes defined on opposite surfaces. We also use this notation to identify the plane and direction of stresses, σ_{ij}, and strains, ε_{ij}.

In the following sections, we will present equilibrium relationships relating stresses to body forces, constitutive relationships relating stresses to strains, and kinematic relationships relating strains to deformations.

8.1.1.1 Equilibrium Equations

In 3D, we have three linear and three rotational equilibrium conditions. To develop these equilibrium conditions, we introduce the term **traction** to refer to stress applied to a surface. Each of the six cube surfaces can experience traction in any of the three directions producing eighteen potentially unique tractions.

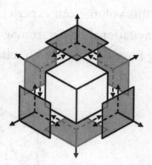

Figure 8.2. The complete set of 18 tractions acting on the surfaces of the infinitesimal cube.

Tractions applied to the surface of the cube parallel to the surface normal generate **normal stresses** (σ_x, σ_y, σ_z) while tractions applied parallel to the surface plane produce **shear stresses** (τ_{xy}, τ_{xz}, τ_{yx}, τ_{yz}, τ_{zx}, τ_{zy}). Tractions come in pairs that act along the same axes, but on opposite surfaces. Rather than defining every traction uniquely, we define paired tractions using internal stresses (σ, τ) and stress differential developed through the cube thickness ($\partial\sigma/\partial x$, $\partial\tau/\partial x$). Since the stress differential is constant through the infinitesimal cube, the change in any stress, $d\sigma_{ij}$, is simply the product of the stress differential, $\partial\sigma_{ij}/\partial x_k$, and the change in dimension, dx_k:

$$d\sigma_{ij} = \frac{\partial\sigma_{ij}}{\partial x_k}dx_k \qquad (8.1)$$

Using this relationship, we define nine pairs of stresses with reference to the cube center. Below, we summarize the three pairs acting on the x-faces of the cube:

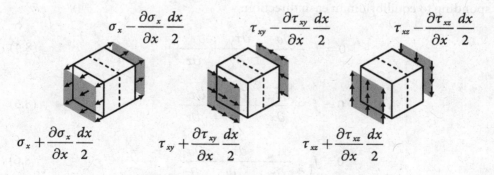

Figure 8.3. Traction couples acting on x-faces.

In addition to tractions, this volume can experience directional **body forces**, f_x, f_y, f_z, usually due to gravitational or electromagnetic fields. To demonstrate the interaction between body forces and applied tractions, consider the 1D case:

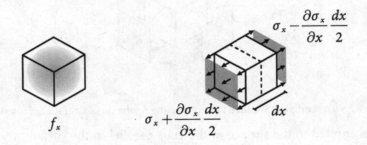

Figure 8.4.   Equilibrium components in 1D.

In 1D, only linear equilibrium along the x-axis is applicable. To express this relationship in consistent units, we multiply the body force by the cube volume, dV, and the applied tractions by their respective surface areas, dA.

$$\sum F_x = 0 = \left(f_x \right) dV + \left(\sigma_x + \frac{\partial \sigma_x}{\partial x} \frac{dx}{2} \right) dA - \left(\sigma_x - \frac{\partial \sigma_x}{\partial x} \frac{dx}{2} \right) dA \qquad (8.2)$$

Gathering terms and using the geometrical equivalence, $dV = dAdx$, we arrive at the simplified expression:

$$0 = f_x + \frac{\partial \sigma_x}{\partial x} \qquad (8.3)$$

In 3D, we follow a similar procedure to extract three parallel relationships corresponding to equilibrium in each direction:

$$0 = f_x + \frac{\partial \sigma_x}{\partial x} + \frac{\partial \tau_{yx}}{\partial y} + \frac{\partial \tau_{zx}}{\partial z} \qquad (8.4)$$

$$0 = f_y + \frac{\partial \tau_{xy}}{\partial x} + \frac{\partial \sigma_y}{\partial y} + \frac{\partial \tau_{zy}}{\partial z} \qquad (8.5)$$

$$0 = f_z + \frac{\partial \tau_{xz}}{\partial x} + \frac{\partial \tau_{yz}}{\partial y} + \frac{\partial \sigma_z}{\partial z} \qquad (8.6)$$

We can summarize the components to these relationships visually:

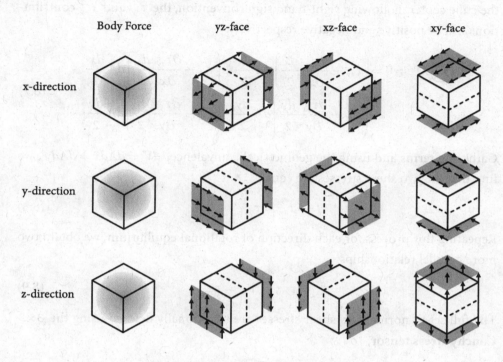

Figure 8.5. Equilibrium components in 3D.

Next, we apply rotational equilibrium conditions, using the cube center as a reference. Rotational equilibrium is only influenced by shear stresses since normal stresses and body forces do not act eccentrically from the cube centroid. Rotational equilibrium about the z-axis only engages two pairs of shear tractions:

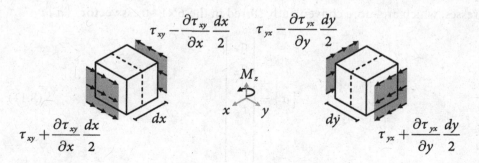

Figure 8.6. Moment equilibrium about the z-axis.

Each traction acts at a distance equal to half of a side length ($dx/2$ or $dy/2$) from the cube center. Following right-hand sign convention, the τ_{xy} and τ_{yx} contributions will be positive and negative respectively:

$$\sum M_z = 0 = \left(\tau_{xy} + \frac{\partial \tau_{xy}}{\partial x} \frac{dx}{2} \right) dA \frac{dx}{2} + \left(\tau_{xy} - \frac{\partial \tau_{xy}}{\partial x} \frac{dx}{2} \right) dA \frac{dx}{2}$$
$$- \left(\tau_{yx} + \frac{\partial \tau_{yx}}{\partial y} \frac{dy}{2} \right) dA \frac{dy}{2} - \left(\tau_{yx} - \frac{\partial \tau_{yx}}{\partial y} \frac{dy}{2} \right) dA \frac{dy}{2} \tag{8.7}$$

Gathering terms and using the geometrical equivalence, $dV = dA dx = dA dy$, we find that the two shear stresses are equal:

$$\tau_{xy} = \tau_{yx} \tag{8.8}$$

Repeating this process for each direction of rotational equilibrium, we obtain two more parallel relationships:

$$\tau_{xz} = \tau_{zx}; \quad \tau_{yz} = \tau_{zy} \tag{8.9}$$

The full set of normal and shear stresses are traditionally defined using the 3×3 **Cauchy stress tensor**, $[\sigma]$:

$$[\sigma] = \begin{bmatrix} \sigma_x & \tau_{xy} & \tau_{xz} \\ \tau_{yx} & \sigma_y & \tau_{yz} \\ \tau_{zx} & \tau_{zy} & \sigma_z \end{bmatrix} \tag{8.10}$$

This matrix implies that there are nine unique stresses; since we have established equivalences between three of the shear stresses, there are actually six unique stresses, which are more conveniently stored in the 6×1 **stress vector**, $\{\sigma\}$:

$$\{\sigma\} = \begin{Bmatrix} \sigma_x \\ \sigma_y \\ \sigma_z \\ \tau_{xy} \\ \tau_{yz} \\ \tau_{xz} \end{Bmatrix} \tag{8.11}$$

8.1.1.2 Constitutive Equations

For derivations in MSA, we relied on the 1D form of Young's law:

$$\sigma = E\varepsilon \tag{8.12}$$

In 3D elasticity, this relationship takes the matrix form:

$$\{\sigma\} = [D]\{\varepsilon\} \tag{8.13}$$

The six unique strains stored in the **strain vector**, $\{\varepsilon\}$, are related to the stresses via the linear elastic **constitutive matrix**, $[D]$:

$$
\begin{Bmatrix} \sigma_x \\ \sigma_y \\ \sigma_z \\ \tau_{xy} \\ \tau_{yz} \\ \tau_{xz} \end{Bmatrix} = \frac{E}{(1+v)(1-2v)}
\begin{bmatrix}
1-v & v & v & 0 & 0 & 0 \\
v & 1-v & v & 0 & 0 & 0 \\
v & v & 1-v & 0 & 0 & 0 \\
0 & 0 & 0 & 0.5-v & 0 & 0 \\
0 & 0 & 0 & 0 & 0.5-v & 0 \\
0 & 0 & 0 & 0 & 0 & 0.5-v
\end{bmatrix}
\begin{Bmatrix} \varepsilon_x \\ \varepsilon_y \\ \varepsilon_z \\ \gamma_{xy} \\ \gamma_{yz} \\ \gamma_{xz} \end{Bmatrix} \tag{8.14}
$$

Poisson's ratio, v, quantifies the magnitude of deformation that occurs transversely to the direction of applied load. Poisson's effect is easy to visualize; if you pull on an elastic band, it will not only get longer, but it will also get thinner. Poisson's ratio is a unitless material property which ranges from 0 to 0.5 for traditional materials; a ratio of 0.5 defines an **incompressible** material.

It is important to recognize that 1D and 2D problems in elasticity must still satisfy this 3D constitutive relationship, a condition typically achieved by assuming that either out-of-plane stresses or strains are negligible. For instance, to achieve the 1D relationship that we used for the truss and beam we set to zero all stresses other than the x-axis normal stress.

$$\sigma_y = \sigma_z = \tau_{xy} = \tau_{yz} = \tau_{xz} = 0 \tag{8.15}$$

These conditions result in simplified expressions for y-axis and z-axis normal strains as well as null shear strains:

$$\varepsilon_y = \varepsilon_z = -v\varepsilon_x; \quad \gamma_{xy} = \gamma_{yz} = \gamma_{xz} = 0 \tag{8.16}$$

Plugging these equations into the general 3D constitutive relationship reduces to the familiar relationship, $\sigma_x = E\varepsilon_x$:

$$\sigma_x = \frac{E}{(1+v)(1-2v)}\left((1-v)\varepsilon_x - v^2\varepsilon_x - v^2\varepsilon_x\right) = \frac{1-v-2v^2}{1-v-2v^2}E\varepsilon_x = E\varepsilon_x \quad (8.17)$$

By neglecting transverse stresses, we assume that the element cross-section is transversely unconstrained, a good approximation for thin elements. If we instead wanted to perform a 1D analysis of a large, thin elastic pad subject to a uniformly-distributed load, we would need to constrain the transverse strains:

$$\varepsilon_y = \varepsilon_z = \gamma_{xy} = \gamma_{yz} = \gamma_{xz} = 0 \quad (8.18)$$

These conditions result in simplified expressions for y-axis and z-axis normal stresses as well as null shear stresses:

$$\sigma_y = \sigma_z = \frac{v}{(1+v)(1-2v)}E\varepsilon_x; \quad \tau_{xy} = \tau_{yz} = \tau_{xz} = 0 \quad (8.19)$$

The primary x-axis normal stress is still proportional to the x-axis normal strain, but uses a slightly more complex coefficient:

$$\sigma_x = \frac{1-v}{(1+v)(1-2v)}E\varepsilon_x \quad (8.20)$$

In 2D, we can develop a parallel pair of relationships corresponding to plane stress and plain strain conditions:

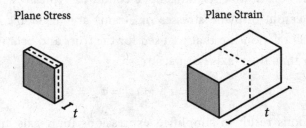

Figure 8.7.   2D plane stress and plane strain.

A **plane stress** condition is frequently assumed in the analysis of thin **membranes** (i.e., a stretched tensile fabric). This condition is achieved when the out-of-plane thickness is small leading to negligible out-of-plane stresses:

$$\sigma_z = \tau_{yz} = \tau_{xz} = 0 \tag{8.21}$$

Algebraic manipulation leads to the following 2D relationship between in-plane strains and stresses:

$$\begin{Bmatrix} \sigma_x \\ \sigma_y \\ \tau_{xy} \end{Bmatrix} = \begin{bmatrix} D^\sigma \end{bmatrix} \begin{Bmatrix} \varepsilon_x \\ \varepsilon_y \\ \gamma_{xy} \end{Bmatrix}; \quad \begin{bmatrix} D^\sigma \end{bmatrix} = \frac{E}{1-v^2} \begin{bmatrix} 1 & v & 0 \\ v & 1 & 0 \\ 0 & 0 & (1-v)/2 \end{bmatrix} \tag{8.22}$$

The out-of-plane normal strain is expressed in terms of in-plane normal strains, while the shear strains are found to be zero:

$$\varepsilon_z = \frac{v}{v-1}(\varepsilon_x + \varepsilon_y); \quad \gamma_{yz} = \gamma_{xz} = 0 \tag{8.23}$$

A **plane strain** condition is frequently assumed in the sectional analysis of long extruded prisms subject to loads orthogonal to the prism axis (i.e., a dam subject to water pressure). This condition is achieved when the out-of-plane thickness is large leading to negligible out-of-plane strains:

$$\varepsilon_z = \gamma_{yz} = \gamma_{xz} = 0 \tag{8.24}$$

Algebraic manipulation leads to the 2D relationship between in-plane strains and stresses:

$$\begin{Bmatrix} \sigma_x \\ \sigma_y \\ \tau_{xy} \end{Bmatrix} = \begin{bmatrix} D^\varepsilon \end{bmatrix} \begin{Bmatrix} \varepsilon_x \\ \varepsilon_y \\ \gamma_{xy} \end{Bmatrix}; \quad \begin{bmatrix} D^\varepsilon \end{bmatrix} = \frac{E}{(1+v)(1-2v)} \begin{bmatrix} 1-v & v & 0 \\ v & 1-v & 0 \\ 0 & 0 & 0.5-v \end{bmatrix} \tag{8.25}$$

The out-of-plane normal stress can be expressed in terms of the primary strains, while the shear strains are zero:

$$\sigma_z = \frac{v}{(1+v)(1-2v)} E(\varepsilon_x + \varepsilon_y); \quad \tau_{yz} = \tau_{xz} = 0 \tag{8.26}$$

8.1.1.3 Kinematic Equations

In our analysis of trusses and beams, we relied on the simple 1D kinematic equation relating axial strain to elongation:

$$\varepsilon = \frac{du}{dx} = \frac{\Delta L}{L} \tag{8.27}$$

In 3D, we refer to the more general form of the **Cauchy strains**, whose components are defined using Einstein notation:

$$\varepsilon_{kl} = \frac{1}{2}\left(\frac{\partial u_l}{\partial x_k} + \frac{\partial u_k}{\partial x_l} \right); \quad k,l \in [1,3] \tag{8.28}$$

The indices in this equation identify three directional components of deformation ($u_1 = u$, $u_2 = v$, $u_3 = w$) and six unique strains. We can expand this expression to obtain the explicit formulation of the strain vector, $\{\varepsilon\}$:

$$\{\varepsilon\} = \begin{Bmatrix} \varepsilon_x \\ \varepsilon_y \\ \varepsilon_z \\ 2\varepsilon_{xy} \\ 2\varepsilon_{yz} \\ 2\varepsilon_{xz} \end{Bmatrix} = \begin{Bmatrix} \varepsilon_x \\ \varepsilon_y \\ \varepsilon_z \\ \gamma_{xy} \\ \gamma_{yz} \\ \gamma_{xz} \end{Bmatrix} = \begin{Bmatrix} \dfrac{\partial u}{\partial x} \\[2mm] \dfrac{\partial v}{\partial y} \\[2mm] \dfrac{\partial w}{\partial z} \\[2mm] \dfrac{\partial u}{\partial y}+\dfrac{\partial v}{\partial x} \\[2mm] \dfrac{\partial v}{\partial z}+\dfrac{\partial w}{\partial y} \\[2mm] \dfrac{\partial u}{\partial z}+\dfrac{\partial w}{\partial x} \end{Bmatrix} = \begin{bmatrix} \dfrac{\partial}{\partial x} & 0 & 0 \\[2mm] 0 & \dfrac{\partial}{\partial y} & 0 \\[2mm] 0 & 0 & \dfrac{\partial}{\partial z} \\[2mm] \dfrac{\partial}{\partial y} & \dfrac{\partial}{\partial x} & 0 \\[2mm] 0 & \dfrac{\partial}{\partial z} & \dfrac{\partial}{\partial y} \\[2mm] \dfrac{\partial}{\partial z} & 0 & \dfrac{\partial}{\partial x} \end{bmatrix} \begin{Bmatrix} u \\ v \\ w \end{Bmatrix} = \left[S^E \right]\{u\} \tag{8.29}$$

In this book, we will use **engineering shear strains** (γ_{xy}, γ_{yz}, γ_{xz}), which are defined to be twice as large as **true shear strains** (ε_{xy}, ε_{yz}, ε_{xz}). It is worth emphasizing that both types of strains are unitless. For simplicity of representation, we gather the partial derivatives in the **elastic partial derivative matrix**, $\left[S^E \right]$.

8.1.2 Boundary Conditions

Having established the governing equations within the BVP domain, we next need to define essential and natural BCs along the boundaries:

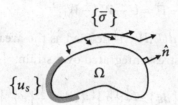

Figure 8.8. Boundary conditions of linear-elastic BVP.

We note that both BCs are now expressed as vectors, rather than scalars, to reflect the inherent directionality of elastic BCs.

Essential BCs in elasticity take the form of prescribed displacements defined along the BVP boundary. Since displacements can be decomposed into directional components, we define the essential BCs using the **prescribed displacement vector**, $\{u_s\}$:

$$\{u_s\} = \begin{Bmatrix} u_s \\ v_s \\ w_s \end{Bmatrix} \tag{8.30}$$

Natural BCs take the form of applied tractions, which are stored in the **applied traction vector**, $\{\bar{\sigma}\}$. Even though the stress state at any BC will be composed of up to six stresses in 3D, the natural BCs are defined using only three directional tractions. The applied traction vector is the product of the Cauchy stress tensor, $[\sigma]$, and the surface normal at the boundary, $\{\hat{n}\}$:

$$\{\bar{\sigma}\} = \begin{Bmatrix} \bar{\sigma}_x \\ \bar{\sigma}_y \\ \bar{\sigma}_z \end{Bmatrix} = \begin{Bmatrix} n_x\sigma_x + n_y\tau_{xy} + n_z\tau_{xz} \\ n_x\tau_{xy} + n_y\sigma_y + n_z\tau_{yz} \\ n_x\tau_{xz} + n_y\tau_{yz} + n_z\sigma_z \end{Bmatrix} = \begin{bmatrix} \sigma_x & \tau_{xy} & \tau_{xz} \\ \tau_{xy} & \sigma_y & \tau_{yz} \\ \tau_{xz} & \tau_{yz} & \sigma_z \end{bmatrix} \begin{Bmatrix} n_x \\ n_y \\ n_z \end{Bmatrix} = [\sigma]\{\hat{n}\} \tag{8.31}$$

It is worth noting that we can mix natural and essential BCs in different axes at any point along the boundary.

8.2 Stiffness Derivation Using the VP

We recall that total potential energy, Π, is defined as the difference between elastic energy, U, and work due to both body forces, W_Ω, and surface tractions, W_Γ:

$$\Pi = U - W_\Omega - W_\Gamma \tag{8.32}$$

The **strain energy density**, dU/dV, is defined as the area under the stress-strain curve, or more explicitly as stress integrated over strain:

$$\frac{dU}{dV} = \int \{\sigma\}^T \{d\varepsilon\} = \frac{1}{2}\{\sigma\}^T \{\varepsilon\} = \frac{1}{2}\{\varepsilon\}^T [D]\{\varepsilon\} \tag{8.33}$$

Introducing the kinematic relationships, we can reformulate this equation using the elastic partial derivative matrix, $\left[S^E\right]$, and the displacement vector, $\{u\}$:

$$\frac{dU}{dV} = \frac{1}{2}\{u\}^T \left[S^E\right]^T [D]\left[S^E\right]\{u\} \tag{8.34}$$

The total strain energy is found by integrating the strain energy density over the domain volume:

$$U = \int_\Omega \frac{dU}{dV} dV = \frac{1}{2}\int_\Omega \{u\}^T \left[S^E\right]^T [D]\left[S^E\right]\{u\} dV \tag{8.35}$$

The work due to internal body forces is the product of displacements and body forces integrated over the volume domain:

$$W_\Omega = \int_\Omega \{u\}^T \{f\} dV \tag{8.36}$$

The work due to external tractions applied at boundaries is the product of displacements and applied tractions integrated over the surface boundary:

$$W_\Gamma = \int_\Gamma \{u\}^T \{\bar{\sigma}\} dA \tag{8.37}$$

Next, we introduce the trial function vector, $\{\tilde{u}\}$:

$$\{\tilde{u}(x, y, z)\} = \sum_{n=1:nnp} \{d_n\} N_n(x, y, z) \tag{8.38}$$

While the form of the solution is familiar, we observe that the index, n, enumerates nodes rather than dofs. The directional components of the trial function and nodal dofs can be expressed using alternate indexing and labelling conventions:

$$\{\tilde{u}\} = \begin{Bmatrix} \tilde{u}_1 \\ \tilde{u}_2 \\ \tilde{u}_3 \end{Bmatrix} = \begin{Bmatrix} \tilde{u} \\ \tilde{v} \\ \tilde{w} \end{Bmatrix}; \quad \{d_n\} = \begin{Bmatrix} d_{xn} \\ d_{yn} \\ d_{zn} \end{Bmatrix} = \begin{Bmatrix} u_n \\ v_n \\ w_n \end{Bmatrix} \tag{8.39}$$

We can express the approximate strain, $\{\tilde{\varepsilon}\}$, by introducing the trial function into the general kinematic relationships we developed earlier:

$$\{\tilde{\varepsilon}\} = \left[S^E\right]\{\tilde{u}\} = \sum_n \left(\left[S^E\right]N_n\right)\{d_n\} = \sum_n \left[B_n^E\right]\{d_n\} \tag{8.40}$$

These kinematic relationships use the **elastic form of the B-matrix**, $\left[B^E\right]$, whose entries are populated using the original B-matrix:

$$\left[B_n^E\right] = \begin{bmatrix} B_{1n} & 0 & 0 \\ 0 & B_{2n} & 0 \\ 0 & 0 & B_{3n} \\ B_{2n} & B_{1n} & 0 \\ 0 & B_{3n} & B_{2n} \\ B_{3n} & 0 & B_{1n} \end{bmatrix}; \quad [B_n] = \begin{bmatrix} B_{1n} \\ B_{2n} \\ B_{3n} \end{bmatrix} = \begin{bmatrix} \dfrac{\partial N_n}{\partial x} \\ \dfrac{\partial N_n}{\partial y} \\ \dfrac{\partial N_n}{\partial z} \end{bmatrix} \tag{8.41}$$

The approximate stress vector is found by introducing the previously-defined strain vector into the general constitutive equation:

$$\{\tilde{\sigma}\} = [D]\{\tilde{\varepsilon}\} = [D]\sum_n \left[B_n^E\right]\{d_n\} \tag{8.42}$$

We are now able to express all of the contributions to the total potential energy:

$$U = \frac{1}{2}\int_\Omega \left(\sum_n \{d_n\}^T \left[B_n^E\right]^T\right)[D]\left(\sum_n \left[B_n^E\right]\{d_n\}\right)dV \tag{8.43}$$

$$W_\Omega + W_\Gamma = \int_\Omega \sum_n \left(\{d_n\}^T \{f\}N_n\right)dV + \int_\Gamma \left(\sum_n \{d_n\}^T \{\bar{\sigma}\}N_n\right)dA \tag{8.44}$$

In order to minimize the total potential energy, we take the partial derivative of the potential with respect to each of the unknown dofs. To maintain concise indexing, we take the partial derivative with respect to the directional vector of unknown nodal dofs:

$$\frac{\partial \Pi}{\partial \{d_m\}} = 0 \qquad (8.45)$$

The derivative rules for vectors parallel those that we established for scaler partial derivatives in the last chapter:

$$\frac{\partial \{\tilde{u}\}}{\partial \{d_m\}} = \frac{\partial}{\partial \{d_m\}}\left(\sum_n \{d_n\} N_n\right) = \sum_n \frac{\partial \{d_n\}}{\partial \{d_m\}} N_n = N_m \qquad (8.46)$$

$$\frac{\partial \{\tilde{\varepsilon}\}}{\partial \{d_m\}} = \frac{\partial}{\partial \{d_m\}}\left(\sum_n \left[B_n^E\right]\{d_n\}\right) = \sum_n \frac{\partial \{d_n\}}{\partial \{d_m\}}\left[B_n^E\right] = \left[B_m^E\right] \qquad (8.47)$$

Using these rules, we are able to obtain the simplified expression for the partial derivatives of the potential:

$$0 = \int_\Omega \sum_n \left[B_m^E\right]^T [D]\left[B_n^E\right]\{d_n\} dV - \int_\Omega \{f\} N_m dV - \int_\Gamma \{\bar{\sigma}\} N_m dA \qquad (8.48)$$

This expression provides all of the ingredients for the global stiffness equation:

$$0 = [K]\{d\} - \{F\} \qquad (8.49)$$

The nodal contributions to the stiffness matrix are $\text{nsd} \times \text{nsd}$ matrices:

$$\left[K_{mn}\right]_{\text{nsd}\times\text{nsd}} = \int_\Omega \left[B_m^E\right]^T [D]\left[B_n^E\right] dV \qquad (8.50)$$

The nodal force and displacement contributions are $\text{nsd} \times 1$ vectors:

$$\left\{F_m\right\}_{\text{nsd}\times 1} = \int_\Omega \{f\} N_m dV + \int_\Gamma \{\bar{\sigma}\} N_m dA; \quad \left\{d_n\right\}_{\text{nsd}\times 1} = \{d_n\} \qquad (8.51)$$

8.2.1 Indexing with Degrees of Freedom

Though we have used nodal indexing thus far in our derivations, we can also define the individual entry contributions using the **global dof indices**, P and Q, corresponding to the ith and j^{th} nodal dof at the mth and n^{th} node respectively:

$$P = (m-1) \times \text{nsd} + i; \quad Q = (n-1) \times \text{nsd} + j \quad (8.52)$$

The nodal and dof indices can be easily extracted from the global indices. We obtain the nodal indices using the **ceiling** operator, $\lceil \ \rceil$, which finds the smallest integer not less than the number specified inside the truncated brackets:

$$m = \lceil P/\text{nsd} \rceil; \quad n = \lceil Q/\text{nsd} \rceil \quad (8.53)$$

The nodal dof can be found using the **modulo** operator, mod, which generates the remainder in the division of two integers:

$$i = \text{mod}(P-1, \text{nsd}) + 1; \quad j = \text{mod}(Q-1, \text{nsd}) + 1 \quad (8.54)$$

As an example, the 12th dof in 3D identifies the z-direction displacement at the 4th node; the 9th dof in 2D identifies the x-direction displacement at the 5th node. Of course, these relationships only apply to homogenous systems; for composite systems, we would need to set up an explicit index to map between global dofs and their corresponding nodes and nodal dofs.

Using these dof indices, we can define the individual entries of the stiffness matrix:

$$K_{PQ} = K_{im,jn} = \int_{\Omega} \left\{ B^E_{m,*i} \right\}^T [D] \left\{ B^E_{n,*j} \right\} dV \quad (8.55)$$

Note that we represent the ith and j^{th} columns of each nodal B-matrix using the subscripts $*i$ and $*j$. The forces and displacements are even easier to express:

$$F_P = F_{im} = \int_{\Omega} f_i N_m dV + \int_{\Gamma} \bar{\sigma}_i N_m dA; \quad d_Q = d_{jn} \quad (8.56)$$

8.2.2 Element Stiffness Equation

By replacing the basis functions with element shape functions and the global domain with the element domain, we obtain the element stiffness equation:

$$\left\{ F^e \right\} = \left[K^e \right]\left\{ d^e \right\} \tag{8.57}$$

The element stiffness matrix is defined as an integral over the element domain:

$$\left[K^e \right] = \int_{\Omega^e} \left[B^E \right]^T \left[D \right]\left[B^E \right] dV \tag{8.58}$$

We can also express the entries of the stiffness matrix explicitly:

$$K_{pq}^e = \int_{\Omega^e} B_{ik}^E D_{kl} B_{jl}^E dV ; \quad k, l \in \left[1 : \text{nsd}(\text{nsd}+1)/2 \right] \tag{8.59}$$

The components of the global stiffness equation are assembled from element contributions following exactly the same procedure as we established for MSA:

$$K_{PQ}^G = \sum K_{PQ}^e ; \quad d_Q^G = d_Q^e ; \quad F_P^G = \sum F_P^e \tag{8.60}$$

8.3 Replicating the Truss Element

To demonstrate the efficacy of the VP for elasticity, we derive the stiffness for a 1D elastic bar with the expectation that it should match the truss formulation we derived in MSA. In 1D, the governing differential equations condense to:

$$\sigma_x = E\varepsilon_x ; \quad \varepsilon_x = \frac{\partial u_x}{\partial x} ; \quad 0 = f_x + \frac{\partial \sigma_x}{\partial x} \tag{8.61}$$

The potential energy in 1D also becomes quite simple:

$$\Pi = \frac{1}{2} \int_{\Omega} E \left(\frac{\partial u_x}{\partial x} \right)^2 dV - \int_{\Omega} u_x f_x dV - \int_{\Gamma} u_x \bar{\sigma}_x dA \tag{8.62}$$

The geometry of the elastic bar in 1D is defined by two points along the x-axis:

Figure 8.9. 1D elastic bar geometry.

The global shape functions match those from the 1D heat rod:

$$N_1^e = \frac{x_2 - x}{x_2 - x_1}; \quad N_2^e = \frac{x - x_1}{x_2 - x_1} \tag{8.63}$$

Assuming that the cross-sectional area remains constant, the element stiffness reduces to a one-dimensional integral over the element length:

$$\left[K^e\right] = \int_{\Omega^e} \left[B^E\right]^T \left[D\right]\left[B^E\right] dV = A \int_L \left[B^E\right]^T \left[D\right]\left[B^E\right] dx \tag{8.64}$$

The elastic B-matrix is trivially found:

$$\left[B^E\right] = \left[\frac{\partial N_1^e}{\partial x} \quad \frac{\partial N_2^e}{\partial x}\right] = \frac{1}{L}\left[-1 \ +1\right] \tag{8.65}$$

The constitutive matrix for a truss is simply the Young's modulus:

$$\left[D\right] = E \tag{8.66}$$

Introducing the constitutive matrix and B-matrix into the element formulation, we obtain a trivial integral:

$$\left[K^e\right] = A\left[B^E\right]^T \left[D\right]\left[B^E\right] \int_L dx = A\left[B^E\right]^T \left[D\right]\left[B^E\right] L \tag{8.67}$$

Upon executing some simple matrix multiplication, we obtain the familiar local stiffness matrix for the truss element:

$$\left[K^e\right] = \frac{EA}{L}\begin{bmatrix} +1 & -1 \\ -1 & +1 \end{bmatrix} \tag{8.68}$$

8.4 2D Triangle Element

Having demonstrated the efficacy of the VP in deriving the 1D truss element, we next turn our attention to the 2D elastic triangle element. By assuming constant thickness, we reduce the element stiffness expression to a 2D area integral:

$$\left[K^e\right] = \int_{\Omega^e} \left[B^E\right]^T \left[D\right]\left[B^E\right] dV = t\int_{A^e} \left[B^E\right]^T \left[D\right]\left[B^E\right] dA \tag{8.69}$$

Next, we transform this integration into the local domain:

$$\left[K^e\right] = t\int_{\bar{A}^e} |J|\left[B^E\right]^T \left[D\right]\left[B^E\right] d\bar{A} \tag{8.70}$$

Consequently, we must map the triangle from the physical to the parent domain:

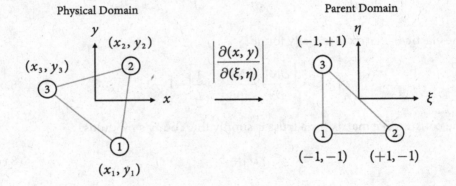

Figure 8.10. Mapping the triangle element from the physical to the parent domain.

The 2D elastic triangle borrows its geometry and shape functions directly from the 2D heat triangle. Hence, we can use the familiar local shape functions:

$$\bar{N}_1^e = \tfrac{1}{2}(-\xi-\eta); \quad \bar{N}_2^e = \tfrac{1}{2}(1+\xi); \quad \bar{N}_3^e = \tfrac{1}{2}(1+\eta) \tag{8.71}$$

As we determined for the heat triangle, the Jacobian, local shape derivative matrix, and B-matrix are constant:

$$\left[\partial \bar{N}\right] = \frac{1}{2}\begin{bmatrix} -1 & 1 & 0 \\ -1 & 0 & 1 \end{bmatrix}; \quad [J] = \frac{1}{2}\begin{bmatrix} x_{21} & y_{21} \\ x_{31} & y_{31} \end{bmatrix}; \quad [B] = \frac{1}{4|J|}\begin{bmatrix} y_{23} & y_{31} & y_{12} \\ x_{32} & x_{13} & x_{21} \end{bmatrix} \tag{8.72}$$

We use the entries of the general B-matrix to populate the elastic B-matrix:

$$
\left[B^E\right] =
\begin{bmatrix}
B_{11} & 0 & B_{12} & 0 & B_{13} & 0 \\
0 & B_{21} & 0 & B_{22} & 0 & B_{23} \\
B_{22} & B_{12} & B_{22} & B_{12} & B_{23} & B_{13}
\end{bmatrix}
= \frac{1}{4|J|}
\begin{bmatrix}
y_{23} & 0 & y_{31} & 0 & y_{12} & 0 \\
0 & x_{32} & 0 & x_{13} & 0 & x_{21} \\
x_{32} & y_{23} & x_{13} & y_{31} & x_{21} & y_{12}
\end{bmatrix}
\tag{8.73}
$$

To maintain consistency between plane stress and plane strain formulations, we use a bit of algebraic manipulation to define one constitutive matrix expressed in terms of an adjusted Young's modulus, $\overline{E}$, and Poisson's ratio, $\overline{v}$:

$$
\left[D\right] = \frac{\overline{E}}{(1-\overline{v}^2)}
\begin{bmatrix}
1 & \overline{v} & 0 \\
\overline{v} & 1 & 0 \\
0 & 0 & (1-\overline{v})/2
\end{bmatrix}
\tag{8.74}
$$

For plane stress problems, we use Young's modulus and Poisson's ratio:

$$
\overline{E}_\varepsilon = E; \quad \overline{v}_\varepsilon = v
\tag{8.75}
$$

For plane strain problems, we adjust the two factors:

$$
\overline{E}_\sigma = \frac{E}{1-v^2}; \quad \overline{v}_\sigma = \frac{v}{1-v}
\tag{8.76}
$$

The reader is encouraged to verify that the adjusted factors for plane strain produce the same constitutive matrix as defined in equation (8.25).

Since all of the entries inside the integral are constant, the only integration required is the trivial area integral in the local domain:

$$
\int_{\overline{A}^e} d\overline{A} = \overline{A}^e = \frac{hw}{2} = \frac{(2)(2)}{2} = 2
\tag{8.77}
$$

The 6×6 elastic triangle element stiffness matrix is thus simply expressed as a matrix multiplication:

$$
\left[K^e\right] = 2t|J|\left[B^E\right]^T \left[D\right]\left[B^E\right]
\tag{8.78}
$$

Although we could find the explicit entries in the elastic triangle stiffness matrix as we did for the heat triangle, it is an unnecessary undertaking. Since we do not need explicit values for code implementation, we leave this exercise to the reader.

8.5 2D Quad Element

Derivation of the elastic quad element closely parallels the derivation we used for the heat quad. The general form of the elastic quad element stiffness in local coordinates exactly matches that of the triangle as expressed in equation (8.70). The 2D elastic quad geometry and shape functions are taken directly from the 2D heat quad leading to an identical mapping strategy:

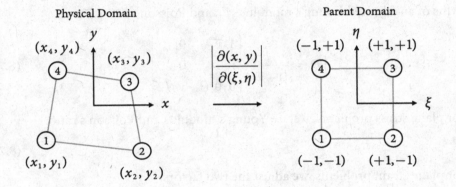

Figure 8.11. Quad mapping from physical to parent domain.

The partial derivative and coordinate matrices are borrowed from the heat quad:

$$[\partial\bar{N}] = \frac{1}{2}\begin{bmatrix} -1+\eta & 1-\eta & 1+\eta & -1-\eta \\ -1+\xi & -1-\xi & 1+\xi & 1-\xi \end{bmatrix}; \quad [x^e] = \begin{bmatrix} x_1 & y_1 \\ x_2 & y_2 \\ x_3 & y_3 \\ x_4 & y_4 \end{bmatrix} \tag{8.79}$$

The Jacobian and general B-matrix are also identical:

$$[J] = [\partial\bar{N}][x^e]; \quad [B] = [J]^{-1}[\partial\bar{N}] \tag{8.80}$$

The elastic B-matrix is populated using the elements of the general B-matrix:

$$[B^E] = \begin{bmatrix} B_{11} & 0 & B_{12} & 0 & B_{13} & 0 & B_{14} & 0 \\ 0 & B_{21} & 0 & B_{22} & 0 & B_{23} & 0 & B_{24} \\ B_{22} & B_{12} & B_{22} & B_{12} & B_{23} & B_{13} & B_{24} & B_{14} \end{bmatrix} \tag{8.81}$$

Since the quad and triangle are both 2D elements, they share the same constitutive matrix (plane stress or plane strain) as defined in equation (8.74).

Because the B-matrix and the Jacobian are not constant, we must perform a nontrivial integration in the element stiffness matrix derivation. As with the heat quad, we transform the integration into a summation using Gaussian quadrature:

$$\left[K^e\right] = t \sum_{j=1:nip} |J(\xi_j, \eta_j)| \left[B^E(\xi_j, \eta_j)\right]^T [D] \left[B^E(\xi_j, \eta_j)\right] w_j \tag{8.82}$$

Elasticity permits greater variation of integration schemes than we explored in heat, but for now the traditional 2×2 integration is still the most reliable choice:

$$nip = 2 \times 2 = 4; \quad (\xi_j, \eta_j) = \left(\pm \frac{1}{\sqrt{3}}, \pm \frac{1}{\sqrt{3}}\right); \quad w_j = 1 \tag{8.83}$$

8.6 Applying Boundary Conditions

The essential and natural BCs for the elastic BVP closely parallel the BCs employed for the heat BVP. Essential BCs are produced by constraining nodal dofs:

$$\left\{\tilde{u}_S(x_n, y_n)\right\} = \left\{u_S(x_n, y_n)\right\} \tag{8.84}$$

Natural BCs are summed from element contributions consisting of body forces integrated over the element domain and applied tractions integrated over external surfaces:

$$\left\{F_m^e\right\} = \int_{\Omega^e} \{f\} N_m^e dV + \int_{\Gamma^e} \{\bar{\sigma}\} N_m^e dA \tag{8.85}$$

For triangle and quad elements, the force contributions are found using the weighting techniques established for heat. Since quadratic applied tractions are common in elasticity, we also want to accommodate a quadratic traction profile:

$$\bar{\sigma}_i = c_{i2} s^2 + c_{i1} s + c_{i0} \tag{8.86}$$

We can obtain the resulting applied traction contribution to the force vector using the quadratic applied flux integration from the previous chapter.

8.7 Stress Recovery

Upon completing the main FEM analysis and obtaining the global and element displacements, we can perform additional post-processing to extract stresses at any point, (ξ, η), within the element:

$$\left\{ \sigma^e(\xi,\eta) \right\} = \left[D^e \right] \left\{ \varepsilon^e(\xi,\eta) \right\} = \left[D^e \right] \left[B^e(\xi,\eta) \right] \left\{ d^e \right\} \tag{8.87}$$

In a triangle element, the B-matrix is constant, so the stresses will not vary within the element domain; we thus only need to report one set of stresses per element. In a quad, the B-matrix and the stresses are not constant; we can thus report the stresses at the centroid, nodes, or integration points of the element. The directional components of the stress vector in 2D are generally defined:

$$\left\{ \sigma(\xi,\eta) \right\} = \begin{Bmatrix} \sigma_x(\xi,\eta) \\ \sigma_y(\xi,\eta) \\ \tau_{xy}(\xi,\eta) \end{Bmatrix} \tag{8.88}$$

For every set of directional stresses, we can use **Mohr's circle** to obtain the **principal stresses**, σ_1 and σ_2, and **principal angle**, θ_p:

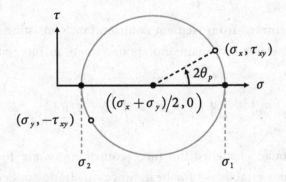

Figure 8.12. Mohr's circle for principal stresses and principal angle in 2D.

The two principal stresses are extracted directly from the geometry of the circle:

$$\sigma_i = \frac{\sigma_x + \sigma_y}{2} \pm \sqrt{\left(\frac{\sigma_x - \sigma_y}{2} \right)^2 + \tau_{xy}^2} \; ; \quad \sigma_1 > \sigma_2 \tag{8.89}$$

The principal angle is also extracted from the geometry of Mohr's circle:

$$\theta_\sigma = \frac{1}{2}\tan^{-1}\left(\frac{2\tau_{xy}}{\sigma_x - \sigma_y}\right) \tag{8.90}$$

An infinitesimal square rotated by the principal angle will experience the two principal stresses void of any shear stresses:

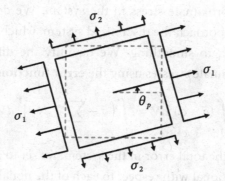

Figure 8.13. Physical manifestation of principal stresses.

We can also extract strains at any point within an element:

$$\left\{\varepsilon^e(\xi,\eta)\right\} = \left[B^e(\xi,\eta)\right]\left\{d^e\right\} \tag{8.91}$$

In determining the principal strains and angle, we note that we typically record shear strains as engineering strains, which are twice the true shear strain; consequently, we need to divide the shear strains in the equation by two:

$$\varepsilon_i = \frac{\varepsilon_x + \varepsilon_y}{2} \pm \sqrt{\left(\frac{\varepsilon_x - \varepsilon_y}{2}\right)^2 + \left(\frac{\gamma_{xy}}{2}\right)^2} \; ; \quad \theta_p = \frac{1}{2}\tan^{-1}\left(\frac{\gamma_{xy}}{\varepsilon_x - \varepsilon_y}\right) \tag{8.92}$$

8.7.1 Stress Smoothing

Because the trial function is composed entirely out of weighted basis functions, the approximate displacements are only C^0 continuous over element boundaries. As a result, stresses and strains, which are derivatives of the displacement field, will be discontinuous over element boundaries. Since these discontinuities are unrealistic, it is common to use a stress smoothing technique to create a continuous stress distribution. The **smoothed stress profile**, $\breve{\sigma}$, closely parallels the structure of the standard trial function:

$$\breve{\sigma}(x, y, z) = \sum_{n=1:nnp} \breve{\sigma}_n N_n(x, y, z) \tag{8.93}$$

Because it is continuous, the smoothed stress profile will not be able to replicate the discontinuous approximate stress in the system. We do, however, want the smoothed stresses to approach the discretized system, which we ensure by using a least-squares approach to smoothing. We quantify the difference between the smoothed and approximate stresses using the **error functional**, Φ:

$$\Phi = \int_\Omega (\tilde{\sigma} - \breve{\sigma})^2 \, dV = \int_\Omega \left(\tilde{\sigma} - \sum \breve{\sigma}_n N_n \right)^2 dV \tag{8.94}$$

In order to minimize the total error in the system, we set to zero the partial derivative of the error functional with respect to each of the nodal dofs:

$$\frac{\partial \Phi}{\partial \breve{\sigma}_m} = 0 = \frac{\partial}{\partial \breve{\sigma}_m} \int_\Omega \left(\tilde{\sigma} - \sum \breve{\sigma}_n N_n \right)^2 dV \tag{8.95}$$

If we take the partial derivative and gather terms, we arrive at the equation:

$$\int_\Omega \tilde{\sigma} N_m dV = \int_\Omega \sum \breve{\sigma}_n N_n N_m dV \tag{8.96}$$

We can represent this relationship as a stiffness equation composed of the **smoothing matrix**, $[H]$, the **discontinuous stress vector**, $\{\tilde{\sigma}\}$, and the **smoothed stress vector**, $\{\breve{\sigma}\}$:

$$\{\tilde{\sigma}\} = [H]\{\breve{\sigma}\} \tag{8.97}$$

The components of these matrices are defined as follows:

$$\tilde{\sigma}_m = \int_\Omega \tilde{\sigma} N_m dV ; \quad H_{mn} = \int_\Omega N_m N_n dV ; \quad \breve{\sigma}_n = \breve{\sigma}_n \tag{8.98}$$

It is common to approximate the H-matrix as a diagonal matrix by summing all of the entries in a single row into the corresponding diagonal entry:

$$H_{nn} = \sum_m \int_\Omega N_m N_n dV = \int_\Omega \left(\sum_m N_m \right) N_n dV = \int_\Omega N_n dV \tag{8.99}$$

This simplification allows us to find the smoothed stress vector without inverting a stiffness matrix:

$$\tilde{\sigma}_n = \frac{\bar{\sigma}_n}{H_n} \tag{8.100}$$

The components of the discontinuous stress vector and the smoothing matrix are assembled from element contributions:

$$\bar{\sigma}_n = \sum_{\forall e} \tilde{\sigma}_n^e ; \quad H_n = \sum_{\forall e} H_n^e \tag{8.101}$$

The element components are defined as integrals over the element domain:

$$\left\{ \tilde{\sigma}^e \right\} = \int_{\Omega^e} \tilde{\sigma} \left\{ N^e \right\} dA ; \quad \left\{ H^e \right\} = \int_{\Omega^e} \left\{ N^e \right\} dA \tag{8.102}$$

For the triangle, this integration is trivial:

$$\left\{ \tilde{\sigma}^e \right\} = \frac{\tilde{\sigma}^e A^e}{3} \left\{ 1 \ 1 \ 1 \right\}^T ; \quad \left\{ H^e \right\} = \frac{A^e}{3} \left\{ 1 \ 1 \ 1 \right\}^T \tag{8.103}$$

For the quad, we can use Gaussian quadrature:

$$\left\{ \tilde{\sigma}^e \right\} = \sum_{j=1:nip} \left| J(\xi_j, \eta_j) \right| \tilde{\sigma}^e(\xi_j, \eta_j) \left\{ \bar{N}^e(\xi_j, \eta_j) \right\} w_j \tag{8.104}$$

$$\left\{ H^e \right\} = \sum_{j=1:nip} \left| J(\xi_j, \eta_j) \right| \left\{ \bar{N}^e(\xi_j, \eta_j) \right\} w_j \tag{8.105}$$

Stress smoothing is a general method that can be equivalently applied to strains and fluxes as long as we have access to the profile of the discontinuous field within each element.

8.8 Code Modifications

A significant portion of the code modifications necessary for implementing the elastic element has already been covered in the last chapter. In our implementation of elastic FEM elements, we rely on several established functions (i.e., genShape and genGauss) and make standard modifications (i.e., changing the defElems function and writing a new element function Ke_elastic). Additionally, we must introduce new functions for generating the elastic B-matrix, calculating principal stresses, and performing stress smoothing.

8.8.1 Modifications to the Element Definition Function

We updated the defElems function by copying the six lines defining the element function name (Ke_elastic) and updating the active dofs in each dimension.

```
% 5. Elasticity
net = net + 1;
kList{net} = 'Ke_elastic';
iad(net,:,1) = [0 0 0 0 0 0 0 0]; % 1D
iad(net,:,2) = [1 1 0 0 0 0 0 0]; % 2D
iad(net,:,3) = [0 0 0 0 0 0 0 0]; % 3D
```

8.8.2 Elastic B-matrix

Though it is a minor operation, we compartmentalize the elastic B-matrix into its own function to make it easier to maintain clean code. We populate the B-matrix, BE, node by node for the number of element nodes, nen (which is defined by the number of columns in the original B-matrix, B).

```
1 function BE = B_elastic(B)
2
3 nen = size(B,2);
4 for j = 1:nen
5   BE(:,2*j-1:2*j) = [B(1,j) 0; 0 B(2,j); B(2,j) B(1,j)];
6 end
```

8.8.3 Elastic Element

The elastic element function, `Ke_elastic`, closely parallels the heat element function from the last chapter. As with the heat element, we use the number of element nodes to distinguish between element types. We also supply the element constitutive matrix, D, to the output field typically reserved for the local element stiffness matrix, ke.

```
1 function [Ke,D,Te] = Ke_elastic(xe,prop)
2
3 nen = size(xe,1);                              % number of element nodes
4 npl = prop(12);                        % number of lambda integration pts
5 npu = prop(13);                            % number of mu integration pts
6
7 % Generate constitutive matrix
8 E = prop(1); v = prop(6); t = prop(7); PS = prop(10);
9 if PS == 1, E = E*(1+2*v)/(1+v)^2; v = v/(1+v); end
10
11 Dl = v*E/((1-2*v)*(1+v))*[1 1 0; 1 1 0; 0 0 0];    % lambda D-matrix
12 Du = E/(2*(1+v))*[2 0 0; 0 2 0; 0 0 1];               % mu D-matrix
13 D = Dl + Du;                                    % complete D-matrix
14
15 % Find element matrix
16 ke = zeros(nen*2);
17
18 if nen == 3
19   [B,detJ] = genShape(xe,[0 0]);
20   BE = B_elastic(B);
21   ke = t*detJ*BE'*D*BE*2;
22 else
23   [ptl,wtl] = genGauss(npl);
24   for i = 1:npl
25     [B,detJ] = genShape(xe,ptl(i,:));
26     BE = B_elastic(B);
27     ke = ke + t*detJ*BE'*Dl*BE*wtl(i);
28   end
29   [ptu,wtu] = genGauss(npu);
30   for i = 1:npu
31     [B,detJ] = genShape(xe,ptu(i,:));
32     BE = B_elastic(B);
33     ke = ke + t*detJ*BE'*Du*BE*wtu(i);
34   end
35 end
36
37 Te = 1;
38 Ke = Te'*ke*Te;
```

The elastic element function differs from the heat function in three significant ways. First, we distinguish between the two types of constitutive matrices using the Boolean variable PS to identify plane stress (PS = 1) and plane strain (PS = 0) conditions.

Second, we use the Gaussian integration loop to generate both the quad and triangle elements. By specifying one integration point (arbitrarily chosen as the origin since both the Jacobian and B-matrix for the triangle are constants) and a weight of two, the summation is reduced to the original triangle formulation:

$$\left[K^e\right] = t \sum_{j=1:1} |J(0,0)| \left[B^E(0,0)\right]^T \left[D\right]\left[B^E(0,0)\right](2) = 2t |J| \left[B^E\right]^T \left[D\right]\left[B^E\right] \quad (8.106)$$

Third, we decompose the constitutive matrix and quad integration into two components corresponding to the first, λ, and second, μ, Lamé components. This adjustment allows us to select different integration schemes to accentuate or diminish certain aspects of the element stiffness. While we have not yet examined the theoretical justifications for this adjustment, we will use the upcoming example to show how selective reduced integration improves the performance of the quad under certain conditions. Note that the two D-matrices, Dl and Du, still sum to produce the standard constitutive matrix as defined in equation (8.74). In order to reproduce the standard elastic matrix, we set both integration point parameters (npl and npu) to the same value (i.e., we would set both npl and npu to 4 integration points for standard 2×2 integration).

8.8.4 Stress and Strain Extraction

Though it is a relatively trivial operation, we also write a simple function for calculating the stresses and strains at a point within an element.

```
1 function [sig,eps] = genStress(de,D,xe,xi)
2
3 if ~exist('xi'), xi = [0 0]; end
4
5 B = genShape(xe,xi);
6 BE = B_elastic(B);
7 eps = BE*de;
8 sig = D*eps;
```

8.8.5 Principal Stresses

Though we will not be using this function directly, we prepare a procedure for generating the principal stresses, sigP, and principal angle, theP, given a set of three stresses stored in a vector, sig.

```
 1 function [sigP,theP] = sigPrinc(sig,strain)
 2
 3 if exist('strain') && strain == 1, sig(3) = sig(3)/2; end
 4
 5 center = (sig(1)+sig(2))/2;
 6 radius = sqrt(0.25*(sig(1)^2+sig(2)^2)+sig(3)^2);
 7 sigP = center + [1 -1]*radius;
 8
 9 if sig(3) == 0, theP = 0;
10 else            theP = atan(2*sig(3)/(sig(1)-sig(2))); end
```

We can also use this function to generate principal strains by setting the Boolean variable strain to 1; this flag changes the shear component to half of the provided shear strain. If the strain variable is not specified or set to anything other than one, the function uses the principal stress equations.

8.8.6 Stress Smoothing

The smtMesh function is effectively a self-contained matrix analysis operation; we generate the element discretized stress and smoothing vectors, assemble the corresponding global vectors, and find the nodal smoothed stresses.

The primary input for this function is the set of discontinuous element stresses, dscElem, which store the stresses at the nodes of each element. The stress distribution inside triangle elements is assumed to be constant; the stress distribution inside quad elements is assumed to be approximately reproduced by multiplying the element stresses at nodes by the nodal shape functions. The discontinuous stress vector, dscNode, and smoothing matrix, H, are assembled from their respective element contributions, dscE and He. Since the stress inside the quad is not constant, we need to also find the stress at the integration points, dscP, in order to obtain the discontinuous element stress contribution. The smoothed stresses are output at global nodes, smtNode, and element nodes, smtElem.

```
 1 function [smtNode,smtElem] = smtMesh(dscElem,xn,ien,nip)
 2
 3 % assume disc is stored as matrix
 4 nel = size(ien,1);                              % number of elements
 5 nen = sum(ien > 0,2);                     % number of element nodes
 6 nnp = size(xn,1);                           % number of nodal points
 7 if ~exist('nip'), nip = 4; end
 8
 9 H = zeros(nnp,1);                               % smoothing matrix
10 dscNode = zeros(nnp,1);                    % discontinuous nodes
11 smtNode = zeros(nnp,1);                        % smoothed nodes
12 smtElem = zeros(nel,nen(1));                   % smoothed elems
13
14 for e = 1:nel
15   % Generate local components
16   xe = xn(ien(e,1:nen(e)),:);
17   if nen(e) == 3
18     [~,detJ,~,~] = genShape(xe,[0 0]);
19     He = detJ*2/3*[1 1 1]';
20     dscE = dscElem(e,1)*detJ*2/3*[1 1 1];
21   else
22     [pts,wts] = genGauss(nip);
23     He = zeros(1,4);
24     dscE = zeros(1,4);
25     for i = 1:nip
26       [~,detJ,N,~] = genShape(xe,pts(i,:));
27       He = He + detJ*N*wts(i);
28       dscP = dscElem(e,:)*N';              % value at integration pt
29       dscE = dscE + dscP*detJ*N*wts(i)';
30     end
31   end
32   % Add element components to global vectors
33   for i = 1:nen(e)
34     H(ien(e,i)) = H(ien(e,i)) + He(i);
35     dscNode(ien(e,i)) = dscNode(ien(e,i)) + dscE(i);
36   end
37 end
38
39 % Generate smoothed node and element components
40 for i = 1:nnp, smtNode(i,:) = dscNode(i,:)/H(i); end
41 for e = 1:nel, smtElem(e,:) = smtNode(ien(e,1:nen(e))); end
```

As stated previously, this function can be used as a general smoothing algorithm for a variety of discontinuous distributions such as heat fluxes, stresses, strains, and stress resultants (which we will cover in the next chapter).

8.9 2D Cantilever Example

In MSA, we analyzed beams using the simplification that the deformation of the beam centerline was sufficient to fully characterize the displacement at every point within the beam. In reality, the underlying assumption that plane sections remain plane is not fully accurate; if we analyze a beam as a 2D elastic problem, we can achieve a more precise representation of beam behavior.

To demonstrate the effectiveness of FEM analysis using triangle and quad elastic elements, we will analyze a 2D cantilever. Since we are modelling the cantilever as a 2D elastic structure, we must be more particular with our BCs. The essential BCs for a 1D beam are produced by restraining both the displacement and the rotation at the support; natural BCs are defined by a concentrated load applied to the free end. There are multiple ways to reproduce these BCs in 2D; we chose a set of BCs consistent with an exact solution originally developed by Love, reformulated by Timoshenko, and verified through FEM analysis by Hughes.

We reproduce the fixed end essential BC by restraining three points in the 2D cantilever. The first point, located at the centroid of the section, is restrained in both the horizontal and vertical directions. The other two points, located at the top and bottom of the section, are only restrained from moving in the horizontal direction; these two points are not vertically restrained since Poisson's effects are likely to produce vertical deformations through the section depth.

The concentrated point load is applied to the free end as a parabolic shear stress state. Since the essential BC only restrains three points, we must also recreate the expected reaction stresses at the fixed end, which we achieve by applying both a parabolic shear profile identical to the one experienced at the free end and a linear horizontal stress.

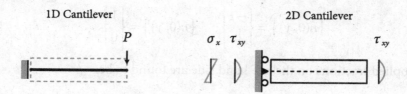

Figure 8.14. Cantilever loaded by a concentrated load as 1D and a 2D problems.

We only model the top half of cantilever in anticipation that the centroid will mark an axis of symmetry; vertical displacements through the section depth will be equal and horizontal displacements will be equal and opposite over the section centroid. In order to reproduce this symmetry, we restrain the nodes lying along the center line from moving in the horizontal direction.

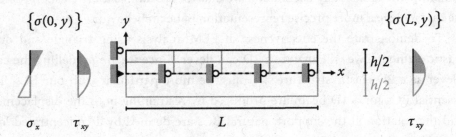

Figure 8.15.   FEM analysis set up.

For this problem, we use the following geometric and material properties:

$$L = 16\,\text{m}; \quad h = 4\,\text{m}; \quad t = 10\,\text{mm}; \quad E = 10\,\text{GPa}; \quad v = 0.3 \qquad (8.107)$$

Since the thickness is small relative to the planar dimensions, a plane stress assumption is used. The applied concentrated load, P, is defined to act downward:

$$P = 10\,\text{kN} \qquad (8.108)$$

The natural BCs are defined by nonzero stresses at the left and right boundaries:

$$\sigma_x(0, y) = \frac{12PL}{th^3}\, y; \quad \tau_{xy}(0, y) = \tau_{xy}(L, y) = \frac{3P}{2th^3}(h^2 - 4y^2) \qquad (8.109)$$

These equations specify a stress state; to generate the applied tractions, we must first find the surface normal at the left-hand and right-hand faces respectively:

$$\{\hat{n}(0, y)\} = \begin{Bmatrix} -1 \\ 0 \end{Bmatrix}; \quad \{\hat{n}(0, y)\} = \begin{Bmatrix} 1 \\ 0 \end{Bmatrix} \qquad (8.110)$$

The applied tractions on the left-hand side are found to be:

$$\{\bar{\sigma}(0, y)\} = [\sigma(0, y)]\{\hat{n}(0, y)\} = \frac{P}{2th^3}\begin{Bmatrix} -24Ly \\ -3(h^2 - 4y^2) \end{Bmatrix} \qquad (8.111)$$

The applied tractions on the right-hand side are similarly obtained:

$$\{\bar{\sigma}(L,y)\} = [\sigma(L,y)]\{\hat{n}(L,y)\} = \frac{P}{2th^3}\begin{Bmatrix} 0 \\ +3(h^2 - 4y^2) \end{Bmatrix} \quad (8.112)$$

The constants defining the x- and y-direction quadratic tractions are then found:

$$\{c_x^0\} = \begin{Bmatrix} 0 \\ -\dfrac{12PL}{th^3} \\ 0 \end{Bmatrix}; \quad \{c_y^0\} = -\{c_y^L\} = \begin{Bmatrix} \dfrac{6P}{th^3} \\ 0 \\ -\dfrac{3P}{2th} \end{Bmatrix} \quad (8.113)$$

The essential BCs are much easier to define than the natural BCs:

$$u(0,0) = v(0,0) = u(0,h/2) = u(x,0) = 0 \quad (8.114)$$

8.9.1 Exact Solution

The method for solving the exact solution to this problem was first developed by Love, then more conventionally described by Timoshenko, and presented in the exact form used in this text by Hughes. The reader is encouraged to verify that this solution satisfies the essential BCs.

$$\begin{Bmatrix} u \\ v \end{Bmatrix} = -\frac{P}{2Eth^3}\begin{Bmatrix} 12x^2 y - 24Lxy + (2+v)(h^2 y - 4y^3) \\ -4x^3 + 12Lx^2 + x(4h^2 + 5vh^2 - 12vy^2) + 12vLy^2 \end{Bmatrix} \quad (8.115)$$

The horizontal deformations at the fixed end are nonzero due to Poisson's effect:

$$u(0,y) = -\frac{P(2+v)}{2Eth^3}(h^2 y - 4y^3) \quad (8.116)$$

The exact stresses are found using the kinematic and constitutive relationships:

$$\{\sigma\} = \begin{Bmatrix} \sigma_x \\ \sigma_y \\ \tau_{xy} \end{Bmatrix} = \frac{P}{2th^3}\begin{Bmatrix} 24(L-x)y \\ 0 \\ 3(4y^2 - h^2) \end{Bmatrix} \quad (8.117)$$

8.9.2 Main Script

The main script for this analysis closely follows the structure we used for the heat example. A significant modification is required to generate the natural BCs; we use the **intForce** function from the previous chapter to calculate the nodal forces based on the nodal coordinates and the vertical element side length (Ls).

```
 1 % exChpt8 - Analysis of a 2D deep cantilever in N and mm base units
 2 clear;
 3
 4 % 0. Convergence inputs
 5 mesh = 3;            % mesh type, 1 = right tri, 2 = left tri, 3 = quad
 6 neX = 2;                          % number of elements along x-axis
 7 neY = neX/2;                      % number of elements along y-axis
 8 npl = 1;                    % number of lambda integration pts
 9 npu = 1;                        % number of mu integration pts
10
11 % 1. Global definitions
12 nsd = 2;                          % number of spatial dimensions
13 ndf = 7;                          % number of degrees of freedom
14 L = 16000; h = 4000; Ls = h/2/neY; t = 10; PS = 0;
15 E = 10000; v = 0.499;  P0 = 10000;
16 cX = [0 -12*P0*L/(t*h^3) 0]; cY = [-6*P0/(t*h^3) 0 3*P0/(2*t*h)];
17
18 % 2. Nodal & 3. Element definitions
19 [xn,ien] = genMesh(L,h/2,neX,neY,nsd,mesh);
20 nnp = size(xn,1);                     % number of nodal points
21 nel = size(ien,1);                    % number of elements
22 idb = zeros(nnp,ndf);                 % index of dofs - supported
23 ds = zeros(nnp,ndf);        % prescribed displacements at supports (mm)
24 Pu = zeros(nnp,ndf);        % applied forces at unrestrained dofs (N)
25 prop = repmat([5 E 0 0 0 0 v t 0 0 PS 0 npl npu 0 0],[nel 1]);
26
27 for n = 1:nnp
28   % Set essential BCs
29   if xn(n,1:2) == [0 h/2], idb(n,1)   = 1; end
30   if xn(n,1:2) == [0   0], idb(n,1:2) = 1; end
31   if xn(n,2)    == 0,      idb(n,1)   = 1; end
32
33   % Set natural BCs
34   if xn(n,1) == 0 || xn(n,1) == L
35     if xn(n,2) ~= 0
36       Pu(n,1) = Pu(n,1) + intForce(xn(n,2)-[Ls 0],cX,t,2);
37       Pu(n,2) = Pu(n,2) + intForce(xn(n,2)-[Ls 0],cY,t,2);
38     end
39     if xn(n,2) ~=  h/2
```

```
40          Pu(n,1) = Pu(n,1) + intForce(xn(n,2)+[0 Ls],cX,t,1);
41          Pu(n,2) = Pu(n,2) + intForce(xn(n,2)+[0 Ls],cY,t,1);
42      end
43      if xn(n,1) == L, Pu(n,1:2) = [0 -Pu(n,2)]; end
44    end
45 end
46
47 % 4. RUN ANALYSIS
48 [results,process] = runAnalysis(Pu,ds,xn,prop,idb,ien);
49 [F,Rs,Fe,Fi,d,du,de] = deal(results{:});
50 [Kuu,Ke,ke,Te,ied,idu,ids] = deal(process{:});
```

We still rely on the genMesh function to generate various meshes. This time, however, we choose to specify twice as many elements in the horizontal dimension to maintain a better aspect ratio. Our analysis relies on three mesh types and three primary mesh refinements:

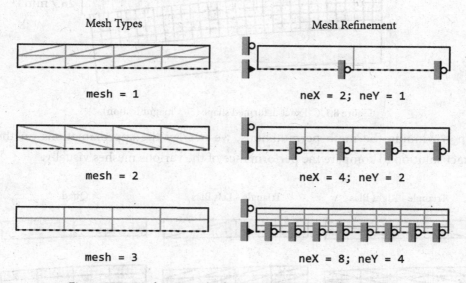

Figure 8.16. Mesh types and refinement for 2D cantilever problem.

8.9.3 Results

The increasing complexity of our analyses is accompanied by an ever-growing plethora of data and results. Though the natural tendency is to capture and present the full set of results, exhaustive tabulations quickly become incomprehensible to human interpretation. It is thus important to highlight the results that motivated the analysis in the first place. For this example, we wish to demonstrate the influence of mesh type and resolution on the overall performance of the FEM approximation with respect to displacements and stresses. Although it is atypical to have access to an exact solution, we take advantage of its availability to set a benchmark for the approximate solutions.

8.9.3.1 Displacements

Using the exact solution from equation (8.115), we draw the exaggerated deformed shape of the cantilever:

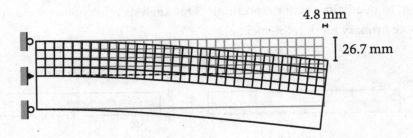

Figure 8.17. Exact deformed shape (50× magnification).

Upon running the mesh permutations, we overlay the approximations on the exact solution to compare the performance of the various meshes visually:

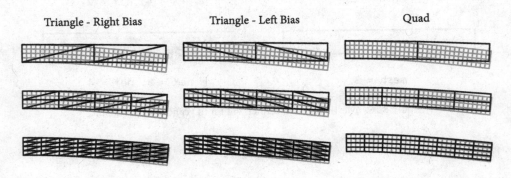

Figure 8.18. Approximate deformed shapes (50× magnification) overlaid on exact solution.

These deformed shapes demonstrate that the approximation generally tends to underestimate the magnitude of deformations. Smaller deformations imply a stiffer structure experiencing smaller stresses; if these results are directly used in design, the resulting system could be under-designed. While underestimation is not an overarching property of FEM analysis, it does highlight the importance of using a suitably fine mesh. We can more precisely analyze the influence of the performance of the various meshes by tabulating the normalized directional displacements at the top-right node. Two additional fine meshes (16x8 and 32x16) are included to better capture the convergence profiles:

nel	Tri - Right (mesh = 1)		Tri - Left (mesh = 2)		Quad (mesh = 3)	
	$\tilde{u}/u$	$\tilde{v}/v$	$\tilde{u}/u$	$\tilde{v}/v$	$\tilde{u}/u$	$\tilde{v}/v$
2x1	0.168	0.192	0.165	0.187	0.390	0.391
4x2	0.436	0.453	0.429	0.442	0.710	0.707
8x4	0.748	0.753	0.741	0.742	0.902	0.899
16x8	0.919	0.919	0.915	0.912	0.972	0.970
32x16	0.977	0.977	0.975	0.973	0.992	0.991

We note that both horizontal and vertical displacements approach the exact solution following a very similar convergence pattern. Also, the two triangle meshes approach the exact solution at nearly identical rates. The vertical displacement in the triangle and quad are thus sufficient to characterize the convergence of the meshes:

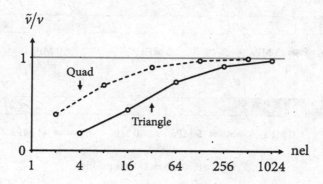

Figure 8.19. Normalized vertical displacement at top-right node.

Even though both curves approach the exact solution asymptotically, the quad element does so much more quickly than does the triangle. Because approximations will never precisely match the exact solution, each analyst must establish an acceptable level of error in the results and then use an appropriate mesh to achieve that level of accuracy.

Since exact solutions are rarely available (if they were, we would not need to use FEM), we must typically rely on convergence to indicate our proximity to an exact solution; if an established element is being used in an appropriate manner, convergence is guaranteed. While convergence studies are incredibly important in FEM analysis, they are underutilized in professional practice. A reliable and easy check when working with commercial FEM software is to subdivide all of the elements in a mesh into four smaller elements and evaluate the analysis. If the results produced by this refined mesh are acceptably close to the original results, then the original mesh was sufficiently refined. If there is a significant change in performance, then the number of elements being used is too low. This subdivision check would thus need to be reiterated with a finer mesh until convergence is demonstrated.

8.9.3.2 Stresses

There are five unique stress distributions (two normal, one shear, and two principal) in 2D elasticity, for which we plot the exact solutions:

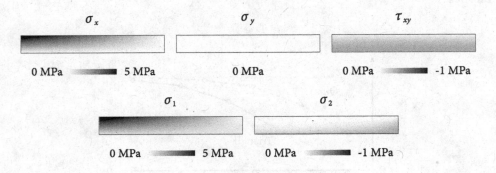

Figure 8.20. Exact stress profiles.

As visually apparent from our analysis, the most dominant stress in the cantilever is the x-direction normal stress. Hence, we will use this stress to evaluate the performance of the approximations. We begin by plotting both the unaltered and smoothed stress profiles of the x-direction stresses:

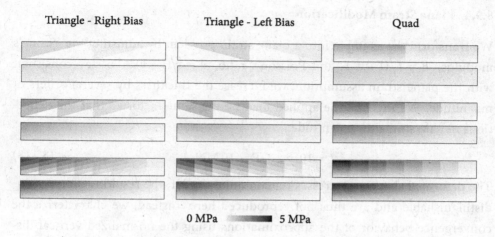

Figure 8.21. X-direction stresses alternatingly unaltered and smoothed.

We note that the coarser meshes demonstrate a clear underestimation of the stresses. In order to evaluate the performance relative to the exact solution, we plot the error distribution in the stress:

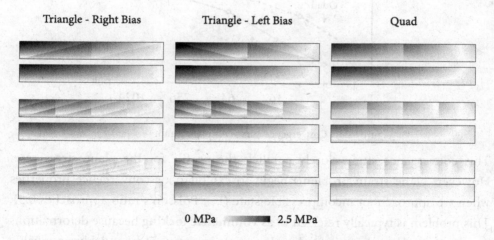

Figure 8.22. X-direction stresses errors alternatingly unaltered and smoothed.

We note that while stress smoothing provides a more convincing overall stress profile, it does not generally reduce the overall magnitude of error in the mesh.

8.9.4 Plane Strain Modification

We transform the example into a plane strain problem by adjusting the Young's modulus, $\bar{E} = E/(1-v^2)$, and Poisson's ratio, $\bar{v} = v/(1-v)$. To stay consistent with the plane strain assumption, we increase the thickness by several orders of magnitude. We increase the applied load by the same factor so that the deflections remain similar in magnitude.

$$t = 10\,\text{m}; \quad P = 10\,\text{MN} \tag{8.118}$$

The deformed shape and stresses remain quite similar; the resulting plots are indistinguishable and are thus not reproduced here. Instead, we characterize the convergence behavior of the approximations using the normalized vertical displacement at the top-right corner of the cantilever:

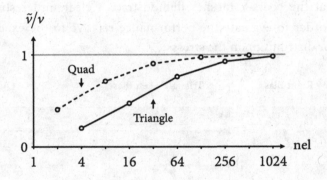

Figure 8.23. Convergence curves for plane strain ($v = 0.3$).

This convergence study closely parallels the profile for the plane stress case. However, plane strain problems begin to exhibit poor convergence properties when we approach an incompressible state (i.e., Poisson's ratio approaches 0.5). This problem is typically referred to as **volumetric locking** because deformations are consistently underestimated and convergence cannot be quickly or reliably achieved. If we execute the plane strain analysis with a Poisson's ratio of 0.499, not only does the overall convergence become unreliable, but the two triangle meshes follow arbitrarily different convergence curves:

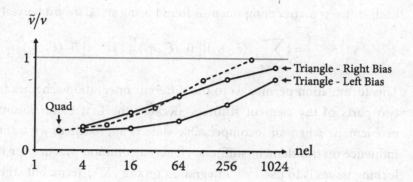

Figure 8.24. Convergence curves for plane strain ($v = 0.499$).

While there is no way to resolve this issue for triangle meshes, we can generate a more reliable quad element using selective reduced integration. As highlighted earlier in the chapter, we have already accommodated for this technique in the element function, `Ke_elastic`.

We begin by splitting up the plane strain constitutive matrix into components corresponding to the **first**, λ, and **second**, μ, **Lamé parameters**:

$$\left[D^{\varepsilon} \right] = \left[D^{\lambda} \right] + \left[D^{\mu} \right] = \begin{bmatrix} \lambda + 2\mu & \lambda & 0 \\ \lambda & \lambda + 2\mu & 0 \\ 0 & 0 & \mu \end{bmatrix} \tag{8.119}$$

The Lamé parameters are not easily interpreted physically, but can be mathematically expressed in terms of Young's modulus and Poisson's ratio:

$$\lambda = \frac{vE}{(1-2v)(1+v)}; \quad \mu = \frac{E}{2(1+v)} \tag{8.120}$$

We observe that the first Lamé parameter distinctly relates to the incompressible state since it becomes very large as Poisson's ratio approaches 0.5. Because the plane stress constitutive matrix is not directly assembled using Lamé parameters, it is not so dramatically affected by Poisson's ratio.

Just as we break up the constitutive matrix into Lamé contributions, we can also break up the element stiffness matrix into two parallel components:

$$\left[K^{\varepsilon} \right] = \left[K^{\lambda} \right] + \left[K^{\mu} \right] \tag{8.121}$$

Each stiffness matrix component is found using standard numerical integration:

$$\left[K^{\lambda,\mu}\right] = t \sum_{j=1:nip} |J(\xi_j, \eta_j)| \left[B^E(\xi_j, \eta_j)\right]^T \left[D^{\lambda,\mu}\right] \left[B^E(\xi_j, \eta_j)\right] w_j \qquad (8.122)$$

This formulation permits us to use different integration schemes to generate the two parts of the element stiffness. Because the first Lamé parameter becomes problematic when an incompressible state is approached, we want to reduce its influence on the element stiffness. Hence, a common practice for resolving these locking issues is to use 1×1 integration on the $\left[K^{\lambda}\right]$ term, but still employ 2×2 integration on the $\left[K^{\mu}\right]$ term. This technique is called **selective reduced integration** and will figure prominently in our discussion of plates and shells. We can also achieve improved results using **uniform reduced integration**. The convergence curves for the full, selectively reduced, and uniformly reduced integration schemes are presented below:

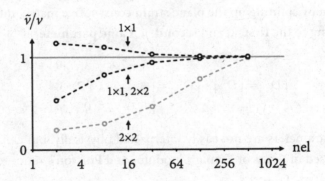

Figure 8.25.   Convergence curves for plane strain ($v = 0.499$) using various integration schemes.

This convergence study clearly demonstrates the improved convergence properties using reduced integration schemes. While uniform reduced integration may appear just as strong as selective reduced integration (and is actually easier to implement), it can lead to singularities in the active stiffness matrix. Since we do not use explicit integration for the triangle, we cannot employ a selective integration scheme; it is thus generally inadvisable to use triangles for plane strain problems.

Chapter 9

Plate and Shell Elements

In MSA, we investigated truss, beam, and frame elements, which resolve loads axially, through bending, and through hybrid axial/bending action, respectively:

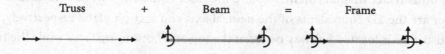

Truss   +   Beam   =   Frame

Figure 9.1. MSA elements.

In our investigation of FEM, we are developing membrane, plate, and shell elements, which are the planar equivalents of the three MSA elements. The membrane resolves loads in plane; the plate resolves loads through bending; and the shell resolves loads both in plane and through bending.

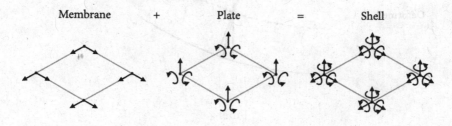

Membrane   +   Plate   =   Shell

Figure 9.2. FEM elements.

The 2D elastic elements we developed in the previous chapter behave as membranes when a plane stress condition is assumed. In this chapter, we will introduce plate and shell elements.

Extending bending theory to plates poses some unique challenges. In this chapter, we evaluate two major bending theories before selecting the theory that permits a reliable FEM implementation. We then present the strong form of the bending BVP and introduce the Variational Principle for bending. We derive the plate quad complete with integration schemes and assemble the shell element from plate and membrane components. Next, we describe the necessary code modifications and conclude with plate and shell examples.

9.1 Two Theories on Bending

It is important to recognize that any beam or plate theory is an idealization of 3D elasticity. We begin our investigation of bending by making the simplification that the displacement of any point within a plate can be completely described by the **midsurface displacement**, w, and the **fiber rotation**, θ. The midsurface and fiber are the 2D equivalents of the neutral axis and section plane respectively; the midsurface is located halfway between the top and bottom surfaces while the fiber is a line segment originally orthogonal to the undeformed midsurface.

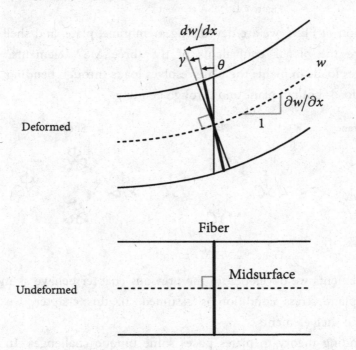

Figure 9.3. Deformed and undeformed section cut of a plate.

We further simplify the plate's bending behavior by requiring that fibers remain straight (a direct extension of the beam assumption that plane sections remain plane), that in-plane deformation at the midsurface is negligible, and that displacements, not just deformations, must be small.

In our investigation of beams, we relied on the assumption that sections not only remain plane but also remain perpendicular to the neutral axis. This assumption only holds true if shear deformations are neglected, leading to a pure bending problem. In general, the fiber's rotation is affected by contributions from both bending and shear. Pure bending is assumed in classic **Euler-Bernoulli beam** theory and in its extension to plates in **Kirchhoff-Love plate** theory. Both of these theories demand that the slope of the midsurface/neutral axis is exactly equal to the fiber/section rotation. The combined effects of shear and bending are considered in the **Timoshenko beam** and **Reissner-Mindlin plate** theories where the fiber/section rotation is found by taking the difference of the midsurface/neutral axis slope and the **shear strain**, γ.

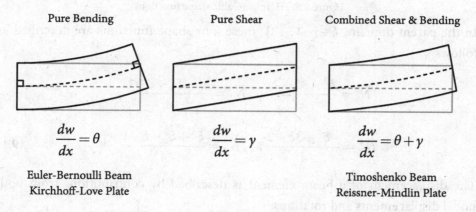

Pure Bending	Pure Shear	Combined Shear & Bending
$\dfrac{dw}{dx} = \theta$	$\dfrac{dw}{dx} = \gamma$	$\dfrac{dw}{dx} = \theta + \gamma$
Euler-Bernoulli Beam Kirchhoff-Love Plate		Timoshenko Beam Reissner-Mindlin Plate

Figure 9.4. Bending and shear deformations in beams and plates.

Pure bending theories are generally accurate for **thin** beams and plates whereas combined bending/shear theories are accurate for **thick** beams and plates. The distinction between thick and thin plates is not precise. Shear effects begin to contribute when the thickness is an order of magnitude lower than the plate's planar dimensions; thus we identify thick plates with a thickness of $t \approx L/10$ and thin plates with a thickness of $t < L/10$. Since the combined bending/shear theories are general, they can also be used for thin plates. In FEM implementation, the shear component, however, leads to locking, which we will investigate further on.

Implementing either theory in FEM technically demands C^1 continuity, which would require that both the out-of-plane displacements and their first derivatives are continuous. For pure bending theories, this requirement is particularly important since the fiber rotation must be exactly equal to the deformed slope. The beam element we used in MSA was based on the Euler-Bernoulli theory and thus maintained C^1 continuity. We achieved this level of continuity by inadvertently using the **Hermite Cubic shape functions** which account for the influence of nodal displacements and rotations:

$$\bar{N}_1^e \qquad\qquad \bar{N}_2^e \qquad\qquad \bar{N}_3^e \qquad\qquad \bar{N}_4^e$$

Figure 9.5. Hermite Cubic shape functions.

In the parent domain, $\xi \in [-1:+1]$, these four shape functions are described as follows:

$$\bar{N}_1^e = \frac{\xi^3 - 3\xi + 2}{4}; \quad \bar{N}_2^e = \frac{\xi^3 - \xi^2 - \xi + 1}{4}$$

$$\bar{N}_3^e = \frac{-\xi^3 + 3\xi + 2}{4}; \quad \bar{N}_4^e = \frac{\xi^3 + \xi^2 - \xi - 1}{4} \tag{9.1}$$

The displacement of a beam element is described by contributions from both nodal displacements and rotations:

$$\tilde{w}^e = w_1 \bar{N}_1^e + \theta_1 \bar{N}_2^e + w_2 \bar{N}_3^e + \theta_2 \bar{N}_4^e \tag{9.2}$$

It is important to note that the Hermite Cubic functions do not follow the conventions and rules of shape functions that we established in Chapter 7; they do not sum to one at every point, nor are they consistently equal to one at their associated node. They do, however, permit us to control the derivative of the displacement:

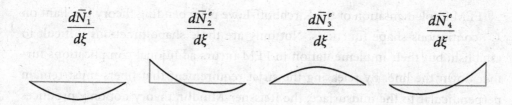

$$\frac{d\bar{N}_1^e}{d\xi} \qquad \frac{d\bar{N}_2^e}{d\xi} \qquad \frac{d\bar{N}_3^e}{d\xi} \qquad \frac{d\bar{N}_4^e}{d\xi}$$

Figure 9.6. Hermite Cubic shape function derivatives.

The shape function derivatives can also be expressed explicitly:

$$\frac{d\bar{N}_1^e}{d\xi} = \frac{3\xi^2 - 3}{4}; \quad \frac{d\bar{N}_2^e}{d\xi} = \frac{3\xi^2 - 2\xi - 1}{4}$$

$$\frac{d\bar{N}_3^e}{d\xi} = \frac{-3\xi^2 + 3}{4}; \quad \frac{d\bar{N}_4^e}{d\xi} = \frac{3\xi^2 + 2\xi - 1}{4} \qquad (9.3)$$

The fiber rotation is defined as the derivative of the vertical displacement or the weighted sum of the shape function derivatives:

$$\tilde{\theta}^e = \frac{d\tilde{w}^e}{d\xi} = w_1 \frac{d\bar{N}_1^e}{d\xi} + \theta_1 \frac{d\bar{N}_2^e}{d\xi} + w_2 \frac{d\bar{N}_3^e}{d\xi} + \theta_2 \frac{d\bar{N}_4^e}{d\xi} \qquad (9.4)$$

The shape functions we used for our investigation of heat and elasticity are classified as **Lagrange shape functions** and only guarantee C^0 continuity. The approximations achieved using Lagrange and Hermite shape functions have different levels of continuity as made visible using a 1D example:

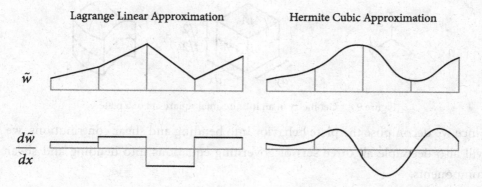

Figure 9.7. Continuity of approximations using Lagrange and Hermite cubic shape functions.

FEM implementation of the Kirchhoff-Love plate bending theory is reliant on C^1 continuous shape functions. Not only are these shape functions difficult to establish, but their implementation in FEM incurs additional complications further down the line. By releasing the strict requirement that fibers must remain perpendicular to the midsurface, the Reissner-Mindlin theory not only produces a more general theory of bending but also facilitates a C^0 FEM implementation. As we will observe in subsequent sections, we can use Lagrangian shape functions to describe the midsurface displacement and fiber rotations independently. Although this approach incurs some theoretical paradoxes, it does generate reliable solutions to planar bending problems.

9.2 Strong Form of the BVP

The strong form of the BVP for the plate is derived from 3D elasticity using the assumptions of the Reissner-Mindlin plate theory. In this section, we present the three governing equations for this theory and outline the strong form of the BVP.

9.2.1 Governing Equations

We present the governing equations for bending categorized by kinematic, constitutive, and equilibrium relationships. We develop these relationships with reference to an infinitesimal square cut of a plate:

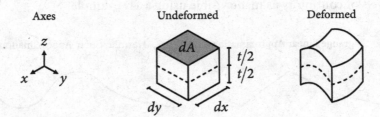

Figure 9.8. Geometry of an infinitesimal square cut of a plate.

Since we decompose the plate behavior into bending and shear contributions, we will also decouple all three sets of governing equations into bending and shear components.

9.2.1.1 Kinematic Equations

In order to assess kinematic relationships, we observe the deformation of the plate using section cuts through the xz-plane and yz-plane:

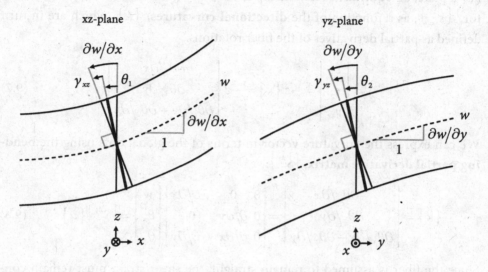

Figure 9.9. Section cuts through a deformed plate in xz-plane and yz-plane.

The Reissner-Mindlin plate theory allows us to fully characterize the plate deformation using the **vertical displacement**, w, and the two **directional fiber rotations**, θ_1 and θ_2. Note that the sign and ordering convention for the fiber rotations is not consistent with the global sign convention:

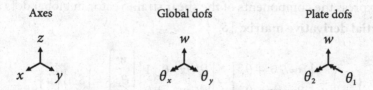

Figure 9.10. Global and plate dofs.

The relationship between plate dofs and global dofs is defined to be:

$$\{u\} = \begin{Bmatrix} w \\ \theta_x \\ \theta_y \end{Bmatrix} = \begin{Bmatrix} w \\ \theta_2 \\ -\theta_1 \end{Bmatrix} \tag{9.5}$$

Through the depth of a beam, bending strains follow a linear profile:

$$\varepsilon = -z\kappa = d\theta/dx \tag{9.6}$$

For a plate, an equivalent relationship is used to express the **bending strain vector**, $\{\varepsilon^M\}$, as a function of the **directional curvatures**, $\{\kappa\}$, which are in turn defined as partial derivatives of the fiber rotations:

$$\{\varepsilon^M\} = \begin{Bmatrix} \varepsilon_x \\ \varepsilon_y \\ \gamma_{xy} \end{Bmatrix} = -z\{\kappa\} = -z\begin{Bmatrix} \partial\theta_1/\partial x \\ \partial\theta_2/\partial y \\ \partial\theta_1/\partial y + \partial\theta_2/\partial x \end{Bmatrix} \tag{9.7}$$

We can express the curvature vector in terms of the global dofs using the **bending partial derivative matrix**, $[S^M]$:

$$\{\kappa\} = \begin{Bmatrix} -\partial\theta_y/\partial x \\ \partial\theta_x/\partial y \\ \partial\theta_x/\partial x - \partial\theta_y/\partial y \end{Bmatrix} = \begin{bmatrix} 0 & 0 & -\partial/\partial x \\ 0 & \partial/\partial y & 0 \\ 0 & \partial/\partial x & -\partial/\partial y \end{bmatrix} \begin{Bmatrix} w \\ \theta_x \\ \theta_y \end{Bmatrix} = [S^M]\{u\} \tag{9.8}$$

Since the fiber is assumed to remain straight, the shear strain must remain constant through the plate thickness. The components of the **shear strain vector**, $\{\varepsilon^V\}$, are thus directly extracted from Figure 9.9:

$$\{\varepsilon^V\} = \begin{Bmatrix} \gamma_{yz} \\ \gamma_{xz} \end{Bmatrix} = \begin{Bmatrix} \partial w/\partial x - \theta_1 \\ \partial w/\partial y - \theta_2 \end{Bmatrix} \tag{9.9}$$

We can express the components of the shear strain vector in global dofs using the **shear partial derivative matrix**, $[S^V]$:

$$\{\varepsilon^V\} = \begin{Bmatrix} \partial w/\partial x + \theta_y \\ \partial w/\partial y - \theta_x \end{Bmatrix} = \begin{bmatrix} \partial/\partial x & 0 & 1 \\ \partial/\partial y & -1 & 0 \end{bmatrix} \begin{Bmatrix} w \\ \theta_x \\ \theta_y \end{Bmatrix} = [S^V]\{u\} \tag{9.10}$$

9.2.1.2 Constitutive Equations

The plate is an elastic body governed by 3D elastic constitutive relationships which relate six strains to six stresses. If we assume that the plate is relatively thin, the out-of-plane normal stress can be neglected ($\sigma_z = 0$), thus allowing us to express the normal vertical strain explicitly:

$$\varepsilon_z = \frac{v}{v-1}(\varepsilon_x + \varepsilon_y) \tag{9.11}$$

Introducing this relationship into the general 3D constitutive equation leads to a reduced constitutive relationship:

$$\begin{Bmatrix} \sigma_x \\ \sigma_y \\ \tau_{xy} \\ \tau_{yz} \\ \tau_{xz} \end{Bmatrix} = \frac{E}{2(1-v^2)} \begin{bmatrix} 2 & 2v & 0 & 0 & 0 \\ 2v & 2 & 0 & 0 & 0 \\ 0 & 0 & 1-v & 0 & 0 \\ 0 & 0 & 0 & 1-v & 0 \\ 0 & 0 & 0 & 0 & 1-v \end{bmatrix} \begin{Bmatrix} \varepsilon_x \\ \varepsilon_y \\ \gamma_{xy} \\ \gamma_{yz} \\ \gamma_{xz} \end{Bmatrix} \tag{9.12}$$

We can break up this relationship into bending and shear components:

$$\{\sigma^M\} = [D^M]\{\varepsilon^M\}; \quad \{\sigma^V\} = [D^V]\{\varepsilon^V\} \tag{9.13}$$

The constitutive bending and shear matrices are defined as follows:

$$[D^M] = \frac{E}{2(1-v^2)} \begin{bmatrix} 2 & 2v & 0 \\ 2v & 2 & 0 \\ 0 & 0 & 1-v \end{bmatrix}; \quad [D^V] = \frac{E}{2(1+v)} \begin{bmatrix} 1 & 0 \\ 0 & 1 \end{bmatrix} \tag{9.14}$$

The two stress vectors correspond to the established strain vectors:

$$\{\sigma^M\} = \begin{Bmatrix} \sigma_x \\ \sigma_y \\ \tau_{xy} \end{Bmatrix}; \quad \{\sigma^V\} = \begin{Bmatrix} \tau_{yz} \\ \tau_{xz} \end{Bmatrix} \tag{9.15}$$

9.2.1.3 Equilibrium Equations

The equilibrium relationships for plates closely parallel the equilibrium equations we have used for beams. Instead of moments and shear, we will use **moment and shear resultants**, which can be interpreted as moments and shears distributed over a unit of length. In this section, we will first define equations relating stresses and resultants and then derive the differential equations governing the relationship between moment and shear resultants.

The relationships between stresses and moment/shear resultants in plates are equivalent to the relationships between stresses and moments/shears in beams. The standard equations for bending and shear stresses in beams are respectively:

$$\sigma = -\frac{Mz}{I}; \quad \tau = \frac{VQ}{Ib} \tag{9.16}$$

Here, Q and I represent the **first and second moments of inertia** respectively:

$$Q = \int z\,dA; \quad I = \int z^2\,dA \tag{9.17}$$

For a rectangular cross-section, these equations can be expressed explicitly:

$$Q = \frac{b}{2}\left(\frac{h^2}{4} - z^2\right); \quad I = \frac{bh^3}{12} \tag{9.18}$$

In a beam with a rectangular cross-section, the bending stresses vary linearly through the section depth while shear stresses vary parabolically.

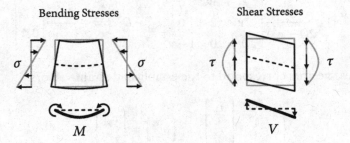

Figure 9.11.   Bending and shear stresses corresponding to positive moments and shears.

At any point along the length of a beam, the moment and shear can be obtained by integrating stresses over the cross-sectional area:

$$M = -\int z\sigma\,dA; \quad V = \int \tau\,dA \tag{9.19}$$

To extend these relationships to plates, we must make two adjustments. First, we analyze the stress contributions over an infinitesimal width, dx or dy, instead of the beam width, b; this modification corresponds to the change from moments and shears to moment and shear resultants.

Second, we need to categorize the stresses depending on whether they produce bending or shear effects. Both normal stresses, σ_x and σ_y, and shear stresses, τ_{xy} and τ_{yx}, will contribute to the moment resultants. These four sets of stresses follow the expected linear profile.

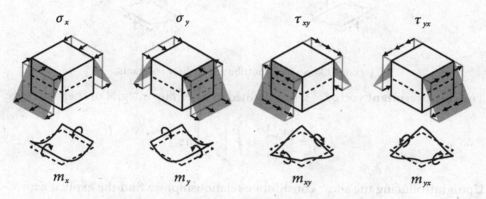

Figure 9.12. Stresses contributing to moment resultants.

Since we know that the two in-plane shear stress profiles are equal, $\tau_{xy} = \tau_{yx}$, the associated moment resultants must also be equal:

$$m_{xy} = m_{yx} \tag{9.20}$$

The **moment resultant vector**, $\{m\}$, is defined as a definite integral of the stresses multiplied by the moment arm relative to the midsurface, z:

$$\{m\} = \begin{Bmatrix} m_x \\ m_y \\ m_{xy} \end{Bmatrix} = -\int\limits_{-t/2}^{+t/2} z \begin{Bmatrix} \sigma_x \\ \sigma_y \\ \tau_{xy} \end{Bmatrix} dz \tag{9.21}$$

Using the established constitutive and kinematic relationships, we obtain an explicit expression for the moment resultant vector:

$$\{m\} = -\int\limits_{-t/2}^{+t/2} z \left[D^M \right] \{\varepsilon^M\} dz = \int\limits_{-t/2}^{+t/2} z^2 \left[D^M \right] \{\kappa\} dz = \frac{t^3}{12} \left[D^M \right] \{\kappa\} \tag{9.22}$$

The remaining two out-of-plane shear stresses contribute to plate shearing:

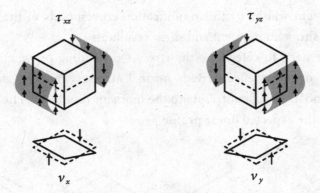

Figure 9.13.   Stresses contributing to shear resultants.

The **shear resultant vector**, $\{v\}$, is defined as a definite integral of shear stresses:

$$\{v\} = \begin{Bmatrix} v_x \\ v_y \end{Bmatrix} = \int_{-t/2}^{+t/2} \begin{Bmatrix} \tau_{xz} \\ \tau_{yz} \end{Bmatrix} dz \qquad (9.23)$$

Upon introducing the shear constitutive relationship, we find the explicit expression for the shear resultant vector:

$$\{v\} = \int_{-t/2}^{+t/2} \left[D^v \right] \left\{ \varepsilon^v \right\} dz = t \left[D^v \right] \left\{ \varepsilon^v \right\} \qquad (9.24)$$

Next, we need to assess the relationships between moment and shear resultants. We begin by assessing moment equilibrium about the x-axis, which is affected by one shear and two moment resultants:

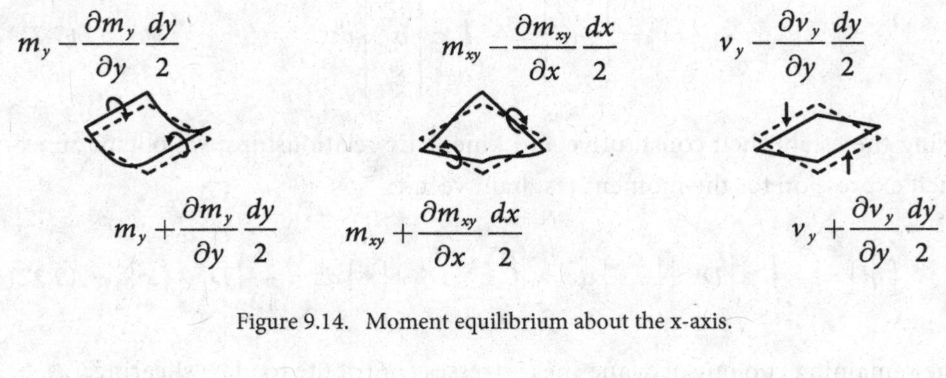

$$m_y - \frac{\partial m_y}{\partial y} \frac{dy}{2} \qquad\qquad m_{xy} - \frac{\partial m_{xy}}{\partial x} \frac{dx}{2} \qquad\qquad v_y - \frac{\partial v_y}{\partial y} \frac{dy}{2}$$

$$m_y + \frac{\partial m_y}{\partial y} \frac{dy}{2} \qquad m_{xy} + \frac{\partial m_{xy}}{\partial x} \frac{dx}{2} \qquad v_y + \frac{\partial v_y}{\partial y} \frac{dy}{2}$$

Figure 9.14.   Moment equilibrium about the x-axis.

We achieve moment equilibrium by summing together the contributions from the moment resultants and the levered contribution from the shear resultant:

$$\sum M_y = 0 = \frac{\partial m_y}{\partial y} dy dx + \frac{\partial m_{xy}}{\partial x} dx dy + v_y dy dx \qquad (9.25)$$

Without executing all of the steps, the moment equilibrium condition reduces to the first equilibrium differential equation for plates:

$$\frac{\partial m_y}{\partial y} + \frac{\partial m_{xy}}{\partial x} + v_y = 0 \qquad (9.26)$$

We repeat these steps for moment equilibrium about the y-axis:

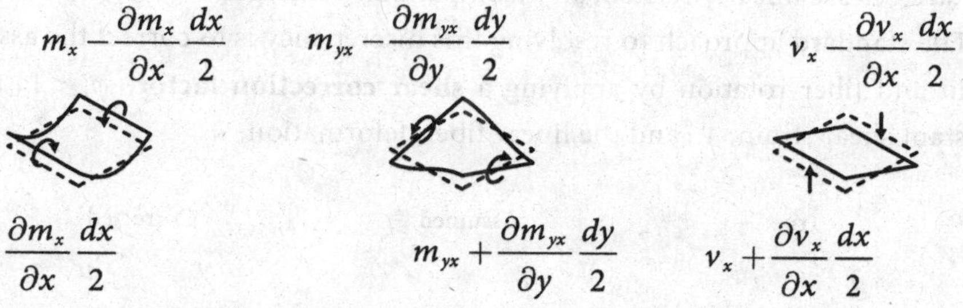

Figure 9.15. Moment equilibrium about the y-axis.

The second equilibrium differential equation for plates is found to be:

$$\frac{\partial m_x}{\partial x} + \frac{\partial m_{yx}}{\partial y} + v_x = 0 \qquad (9.27)$$

Lastly, we look at vertical equilibrium, including body forces:

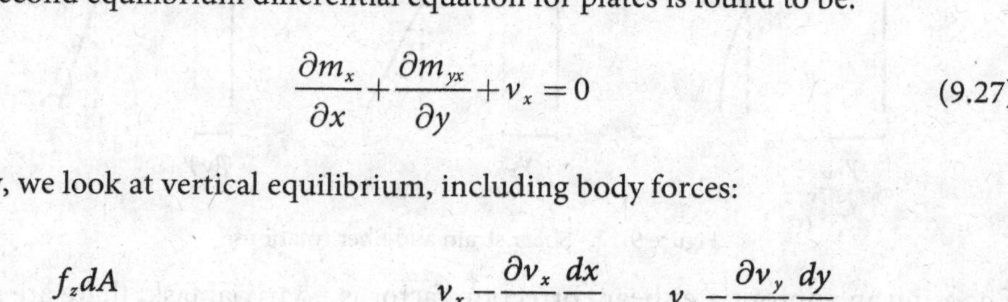

Figure 9.16. Vertical equilibrium.

The third and final differential equation of equilibrium is found to be:

$$\frac{\partial v_x}{\partial x} + \frac{\partial v_y}{\partial y} + f_z = 0 \tag{9.28}$$

9.2.1.4 Shear Correction Factor

A keen reader may note that the shear strains and stresses are incompatible. In our derivation of equilibrium relationships, we assumed that shear stresses followed a parabolic distribution through the thickness. In our kinematic investigation, however, we demanded that the fiber remains straight, implying that the shear strain is constant through the plate thickness. Since shear strains and stresses are linearly related, an incompatibility emerges.

The standard approach to resolving this discrepancy is to correct the assumed strain and fiber rotation by applying a **shear correction factor**, $\varphi_s < 1$, to the constant shear strain, γ, and the linear fiber deformation, u:

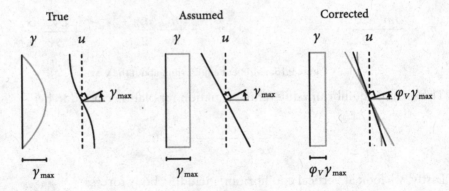

Figure 9.17. Shear strain and fiber rotations.

Choosing an appropriate shear correction factor is no trivial task; there are many publications advocating various approaches for its derivation. For an isotropic, homogenous section, Timoshenko recommends using the following:

$$\varphi_v = \frac{10(1+v)}{12+11v} \approx \frac{5}{6} \tag{9.29}$$

Perhaps counterintuitively, the shear factor is not typically used to redefine the shear strain. Instead, we correct the shear stress vector:

$$\{\sigma^V\} = \varphi_V \left[D^V\right]\{\varepsilon^V\} \tag{9.30}$$

Consequently, the shear resultant vector is also corrected:

$$\{v\} = \varphi_V t \left[D^V\right]\{\varepsilon^V\} \tag{9.31}$$

9.2.2 Boundary Conditions

Since the plate is a simplified version of 3D elasticity, we can technically express the BCs using 3D stresses and displacements. It is, however, more convenient to express the BCs using terms specific to plate theory, namely moment/shear resultants, fiber rotations, and the midsurface displacement. Since we know that the plate will lie in the xy-plane, the BVP is defined by 2D geometry and 3D displacements; thus, we visualize the BVP in 3D using an isometric projection:

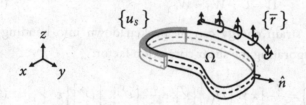

Figure 9.18. Plate BVP represented using isometric projection.

The essential BCs are expressed using the prescribed displacement vector, $\{u_s\}$, composed of the vertical displacement, w_s, and the two directional rotations, θ_{xs} and θ_{ys}:

$$\{u_s\} = \begin{Bmatrix} w_s \\ \theta_{xs} \\ \theta_{ys} \end{Bmatrix}. \tag{9.32}$$

Note that these prescribed displacements are presented using global rather than plate sign convention.

The natural BCs are expressed in terms of the **applied stress resultants vector,** $\{\bar{r}\}$, which stores contributions from the shear and moment resultants:

$$\{\bar{r}\} = \begin{Bmatrix} \bar{v}_z \\ \bar{m}_x \\ \bar{m}_y \end{Bmatrix} = \begin{Bmatrix} v_{xz}n_x + v_{yz}n_y \\ m_x n_x + m_{xy}n_y \\ m_{xy}n_x + m_y n_y \end{Bmatrix} = \begin{bmatrix} v_{xz} & v_{yz} \\ m_x & m_{xy} \\ m_{xy} & m_y \end{bmatrix} \begin{Bmatrix} n_x \\ n_y \end{Bmatrix} \tag{9.33}$$

Because the plate lies in the xy-plane, the boundary normal, $\{\hat{n}\}$, is reduced to two entries:

$$\{\hat{n}\} = \begin{Bmatrix} n_x \\ n_y \end{Bmatrix} \tag{9.34}$$

9.3 Stiffness Derivation Using the VP

The Variational Principle for the plate is derived using the familiar principle of least action. We begin by recalling the general form of potential energy:

$$\Pi = U - W_\Omega - W_\Gamma \tag{9.35}$$

The volumetric internal strain energy can be broken down into bending and shear contributions incorporating the shear correction factor:

$$\frac{dU}{dV} = \frac{1}{2}\{\varepsilon\}^T\{\sigma\} = \frac{z^2}{2}\{\kappa\}^T\left[D^M\right]\{\kappa\} + \frac{\varphi_s}{2}\{\varepsilon^s\}^T\left[D^s\right]\{\varepsilon^s\} \tag{9.36}$$

The internal strain energy is defined as a volume integral of the strain energy density over the domain, which can be decomposed into nested integrals over the area and through the thickness:

$$U = \int_\Omega \frac{dU}{dV} dV = \int_A \left(\int_{-t/2}^{+t/2} \frac{dU}{dV} dz \right) dA \tag{9.37}$$

We first resolve the integral through the plate thickness:

$$\int_{-t/2}^{+t/2} \frac{dU}{dV} dz = \frac{1}{2}\{\kappa\}^T\left[D^M\right]\{\kappa\}\left(\int_{-t/2}^{+t/2} z^2 dz\right) + \frac{\varphi_v}{2}\{\varepsilon^v\}^T\left[D^v\right]\{\varepsilon^v\}\left(\int_{-t/2}^{+t/2} dz\right) \tag{9.38}$$

Upon performing the trivial integration, we obtain the expression for the total internal strain energy composed of bending, M, and shear, V, contributions:

$$U = U^M + U^V = \frac{t^3}{24} \int_A \{\kappa\}^T [D^M] \{\kappa\} dA + \frac{\varphi_V t}{2} \int_A \{\varepsilon^V\}^T [D^V] \{\varepsilon^V\} dA \quad (9.39)$$

There are multiple ways of applying external work to the system, but we will limit our discussion to those mechanisms most likely to occur in practical applications. For the internal work performed on the system by body forces, we could technically consider both moment and shear resultants. However, moments distributed over an area are practically nonexistent, so we limit internal work to gravity-induced vertical forces:

$$W_\Omega = \int_\Omega f_z w \, dV = t \int_A f_z w \, dA \quad (9.40)$$

We note that the body force can also be expressed as an applied force distributed over area, $p_z = t f_z$.

The external work on the system can be performed by both concentrated and distributed loads:

$$W_\Gamma = \int_\Gamma \{u\}^T \{\bar{r}\} ds + \sum_{\forall P} \{u\}^T \{P\} \quad (9.41)$$

We have already established the applied stress resultants vector, $\{\bar{r}\}$; we now define the applied force vector, $\{P\}$, which stores applied concentrated forces and moments:

$$\{P\} = \begin{Bmatrix} F_z \\ M_x \\ M_y \end{Bmatrix} \quad (9.42)$$

Having defined the potential, we next introduce the trial function:

$$\{\tilde{u}(x, y, z)\} = \sum_{n=1:\text{nnp}} \{d_n\} N_n(x, y, z) \quad (9.43)$$

The approximate displacement vector and nodal dof vector are defined by one vertical displacement and two rotations each:

$$\{\tilde{u}\} = \begin{Bmatrix} \tilde{w} \\ \tilde{\theta}_x \\ \tilde{\theta}_y \end{Bmatrix}; \quad \{d_n\} = \begin{Bmatrix} w_n \\ \theta_{xn} \\ \theta_{yn} \end{Bmatrix} \tag{9.44}$$

We next define the approximate curvature, $\{\tilde{\kappa}\}$, and shear strain, $\{\tilde{\varepsilon}^V\}$, vectors:

$$\{\tilde{\kappa}\} = \begin{bmatrix} S^M \end{bmatrix}\{\tilde{u}\} = \sum_n \left(\begin{bmatrix} S^M \end{bmatrix} N_n \right)\{d_n\} = \sum_n \begin{bmatrix} B_n^M \end{bmatrix}\{d_n\} \tag{9.45}$$

$$\{\tilde{\varepsilon}^V\} = \begin{bmatrix} S^V \end{bmatrix}\{\tilde{u}\} = \sum_n \left(\begin{bmatrix} S^V \end{bmatrix} N_n \right)\{d_n\} = \sum_n \begin{bmatrix} B_n^V \end{bmatrix}\{d_n\} \tag{9.46}$$

These expressions use new **bending**, $\begin{bmatrix} B^M \end{bmatrix}$, and **shear**, $\begin{bmatrix} B^V \end{bmatrix}$, **B-matrices** with nodal contributions defined as follows:

$$\begin{bmatrix} B_n^M \end{bmatrix} = \begin{bmatrix} 0 & 0 & -B_{1n} \\ 0 & B_{2n} & 0 \\ 0 & B_{1n} & -B_{2n} \end{bmatrix}; \quad \begin{bmatrix} B_n^V \end{bmatrix} = \begin{bmatrix} B_{1n} & 0 & N_n \\ B_{2n} & -N_n & 0 \end{bmatrix} \tag{9.47}$$

The entries are populated using the general B-matrix derived in Chapter 7.

$$\begin{bmatrix} B_n \end{bmatrix} = \begin{bmatrix} B_{1n} \\ B_{2n} \end{bmatrix} = \begin{bmatrix} \dfrac{\partial N_n}{\partial x} \\ \dfrac{\partial N_n}{\partial y} \end{bmatrix} \tag{9.48}$$

With these approximations in hand, we write out the approximate internal strain energy contributions from bending and shear respectively:

$$U^M = \frac{t^3}{24} \int_A \left(\sum_n \{d_n\}^T \begin{bmatrix} B_n^M \end{bmatrix}^T \right) \begin{bmatrix} D^M \end{bmatrix} \left(\sum_n \begin{bmatrix} B_n^M \end{bmatrix}\{d_n\} \right) dA \tag{9.49}$$

$$U^V = \frac{\varphi_V t}{2} \int_A \left(\sum_n \{d_n\}^T \begin{bmatrix} B_n^V \end{bmatrix}^T \right) \begin{bmatrix} D^V \end{bmatrix} \left(\sum_n \begin{bmatrix} B_n^V \end{bmatrix}\{d_n\} \right) dA \tag{9.50}$$

We also present the work due to internal body forces and applied forces:

$$W_\Omega = \int_A tf_z \left(\sum_n w_n N_n \right) dA \tag{9.51}$$

$$W_\Gamma = \int_\Gamma \{\bar{r}\}^T \left[\sum_n \{d_n\}[N_n] \right] ds + \sum_{\forall P} \{P\}^T \left[\sum_n \{d_n\}[N_n] \right] \tag{9.52}$$

Guided by the Variational Principle, we take the familiar set of partial derivatives:

$$\frac{\partial \Pi}{\partial \{d_m\}} = 0 \tag{9.53}$$

We also establish the partial derivatives of the approximate displacements, curvatures, and shear strains with respect to the dofs:

$$\frac{\partial \{\tilde{u}\}}{\partial \{d_m\}} = N_m; \quad \frac{\partial \{\tilde{\kappa}\}}{\partial \{d_m\}} = \left[B_m^M \right]; \quad \frac{\partial \{\tilde{\varepsilon}^V\}}{\partial \{d_m\}} = \left[B_m^V \right] \tag{9.54}$$

Using these relationships, we find the partial derivatives of the potential:

$$\frac{\partial U^M}{\partial \{d_m\}} = \frac{t^3}{12} \int_A \sum_n \left[B_m^M \right]^T \left[D^M \right] \left[B_n^M \right] \{d_n\} dA \tag{9.55}$$

$$\frac{\partial U^V}{\partial \{d_m\}} = \varphi_V t \int_A \sum_n \left[B_m^V \right]^T \left[D^V \right] \left[B_n^V \right] \{d_n\} dA \tag{9.56}$$

$$\frac{\partial W}{\partial \{d_m\}} = \int_A t \{0 \ 0 \ f_z\}^T N_m dA + \int_\Gamma \{\bar{r}\}^T N_m ds + \sum_{\forall P} \{P\}^T N_m \tag{9.57}$$

These equations provide all of the components of the global stiffness equation:

$$0 = [K]\{d\} - \{F\} \tag{9.58}$$

The nodal contributions to the stiffness matrix are 3×3 matrices:

$$[K_{mn}]_{3\times3} = \frac{t^3}{12} \int_A \left[B_m^M \right]^T \left[D^M \right] \left[B_n^M \right] dA + \varphi_V t \int_A \left[B_m^V \right]^T \left[D^V \right] \left[B_n^V \right] dA \tag{9.59}$$

The nodal force and displacement contributions are 3×1 vectors:

$$\left\{ d_n \right\}_{3\times1} = \left\{ d_n \right\} \tag{9.60}$$

$$\left\{ F_m \right\}_{3\times1} = \int_A t \left\{ f_z \right\} N_m dA + \int_\Gamma \left\{ \overline{r} \right\}^T N_m ds + \sum_{\forall P} \left\{ P \right\}^T N_m \tag{9.61}$$

As with membranes, the matrix components are presented using nodal numbering, m and n; using the indexing techniques presented in the past chapter, we can rewrite the contributions using the global dof indices, P and Q.

The element stiffness equation and the BCs can be derived following the procedures established in the last two chapters; both exercises are left to the reader. There are, however, two essential BCs specific to plates that we want to translate precisely from the continuous BVP to the discretized FEM implementation:

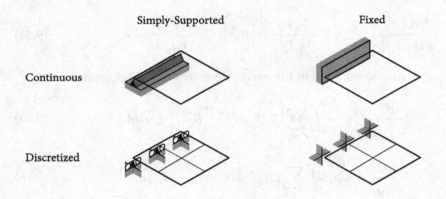

Figure 9.19. Common essential BCs transformed from continuous to discretized.

Transformation of the simply-supported condition requires that only the rotation along the plate edge is released; both the vertical displacement and the rotation orthogonal to the plate edge are restrained. At the corner of simply-supported plates, both rotations should be restrained. Transformation of the fixed condition is achieved by simply restraining the vertical displacement and both rotations.

9.4 Plate Quad Element

Since the plate can only be analyzed in 2D, we are limited to either a triangle or a quad element. We begin with the element stiffness matrix expressed as a local area integral:

$$\left[K^e\right] = \frac{t^3}{12} \int_{\overline{A}^e} |J| \left[B^M\right]^T \left[D^M\right] \left[B^M\right] d\overline{A} + \varphi_V t \int_{\overline{A}^e} |J| \left[B^V\right]^T \left[D^V\right] \left[B^V\right] d\overline{A} \quad (9.62)$$

As thickness becomes smaller, Kirchhoff-Love's thin plate theory dictates that the shear strain should become negligible. In our stiffness formulation, however, the bending component decreases quickly at a cubic rate while the shear component decreases much more slowly at a linear rate. Hence, our FEM implementation disproportionally represents the shear component for thin plates. This phenomenon leads to **shear locking** and thus poses a significant issue for thin plate elements. As we observed for plane strain elasticity, a convenient technique for resolving locking issues in FEM is to employ a selective reduced integration scheme. Since triangular elements do not benefit from selective reduced integration schemes, we automatically limit our discussion to the quad element.

The quad plate element follows the same mapping techniques as used for the 2D heat and elasticity elements. Hence, the Jacobian, partial derivative, and B-matrices are defined as before.

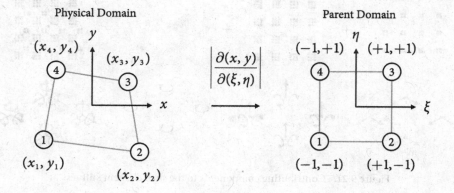

Figure 9.20. Quad mapping from physical to parent domain.

The element stiffness matrix is found using contributions from the bending and shear stiffness matrices, which are found using two numerical integrations:

$$\left[K^e \right] = \left[K^M \right] + \left[K^V \right] \tag{9.63}$$

$$\left[K^M \right] = \frac{t^3}{12} \sum_{j=1:nip} \left| J(\xi_j, \eta_j) \right| \left[B^M(\xi_j, \eta_j) \right]^T \left[D^M \right] \left[B^M(\xi_j, \eta_j) \right] w_j \tag{9.64}$$

$$\left[K^V \right] = \varphi_V t \sum_{j=1:nip} \left| J(\xi_j, \eta_j) \right| \left[B^V(\xi_j, \eta_j) \right]^T \left[D^V \right] \left[B^V(\xi_j, \eta_j) \right] w_j \tag{9.65}$$

To combat shear locking, it is common practice to use 2×2 integration on the bending component and 1×1 on the shear component. Such selective reduced integration reduces shear stiffness relative to bending stiffness; we will demonstrate the efficacy of this integration scheme in the upcoming plate example.

9.5 Shell Quad Element

The shell quad element is assembled from the plate and membrane elements just like the frame element is composed of entries from the beam and truss elements.

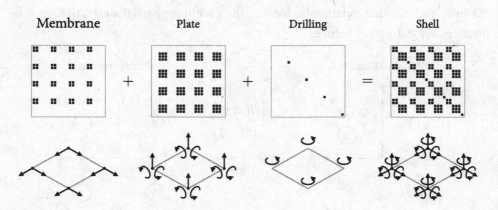

Figure 9.21. Contributing components to the shell element stiffness.

Since the shell element will be rotated into 3D, it should be stable in all six dofs. With the membrane contributing two dofs and the plate contributing three dofs,

we are one dof short of a stable element. In order to achieve stability, we introduce a fictitious **drilling stiffness**, $\left[K^{drilling} \right]$:

$$\left[K^{shell} \right] = \left[K^{membrane} \right] + \left[K^{plate} \right] + \left[K^{drilling} \right] \tag{9.66}$$

This matrix is produced by multiplying the identity matrix by a drilling stiffness
factor. Bathe recommends setting this value to one-thousandth of the minimum,
nonzero entry of the diagonal of the plate and membrane elements:

$$\left[K^{drilling} \right] = \frac{\min \left| K_{ii}^{membrane}, K_{jj}^{plate} \right|}{1000} \left[I \right] \tag{9.67}$$

9.5.1 Rotating the Shell

The shell element is primarily used in 3D, but the plate and membrane stiffnesses
are derived in 2D. Thus, we need to rotate the shell geometry into local 2D coordinates and then transform the stiffness matrix into global 3D coordinates.

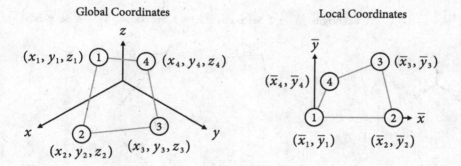

Figure 9.22. Shell geometry in the global and local coordinate systems.

The translation from global to local coordinates can be achieved in four steps.
First, we adjust the global coordinates of the shell $\left\{ x^{o} \right\}$ so that the coordinates of
the first node are consistent with the origin:

$$\begin{Bmatrix} x_i^o \\ y_i^o \\ z_i^o \end{Bmatrix} = \begin{Bmatrix} x_i - x_1 \\ y_i - y_1 \\ z_i - z_1 \end{Bmatrix} \tag{9.68}$$

Second, we place the local x-axis at the first node oriented to the second node:

$$\{n_x\} = \{x_2^o\} / |x_2^o|$$

(9.69)

Third, we define the z-axis by taking the cross product of the local x-axis and any point in the plane of the element (i.e., the third node). Following the right-hand rule, the z-axis should point out of the page in the local coordinate system:

$$\{n_z\} = \frac{\{n_x\} \times \{x_3^o\}}{|\{n_x\} \times \{x_3^o\}|}$$

(9.70)

Finally, we find the local y-axis by taking the cross product of the z-axis and the x-axis once again guided by the right-hand rule:

$$\{n_y\} = \frac{\{n_z\} \times \{n_x\}}{|\{n_z\} \times \{n_x\}|}$$

(9.71)

We visualize these steps in the global and local coordinate systems:

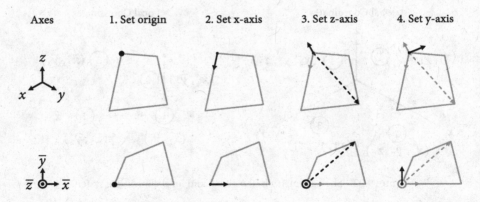

Figure 9.23. Steps involved in rotating the shell element visualized in global and local coordinates.

We conclude by populating the rotation matrix using the direction cosines:

$$[Q] = [\{n_x\} \; \{n_y\} \; \{n_z\}] = \begin{bmatrix} n_{x1} & n_{y1} & n_{z1} \\ n_{x2} & n_{y2} & n_{z2} \\ n_{x3} & n_{y3} & n_{z3} \end{bmatrix}$$

(9.72)

The local coordinates are found using this rotation matrix:

$$\begin{Bmatrix} \bar{x}_i \\ \bar{y}_i \\ \bar{z}_i \end{Bmatrix} = [Q] \begin{Bmatrix} x_i^o \\ y_i^o \\ z_i^o \end{Bmatrix} \tag{9.73}$$

If we have implemented the algorithm correctly and used a planar quad, the local z-coordinate, $\bar{z}_i$, for all nodes should be zero.

Having established the rotation matrix, we define the global shell stiffness matrix using the established MSA transformation formula:

$$\left[K^e \right]_{24 \times 24} = \left[T^e \right]_{24 \times 24}^T \left[k^e \right]_{24 \times 24} \left[T^e \right]_{24 \times 24} \tag{9.74}$$

Since the shell element has four nodes which may each experience a full set of displacements and rotations, the transformation matrix is assembled from eight rotation matrices:

$$\left[T^e \right]_{24 \times 24} = \begin{bmatrix} [Q]_{3 \times 3} & & & & & & & \\ & [Q]_{3 \times 3} & & & & & & \\ & & [Q]_{3 \times 3} & & & & & \\ & & & [Q]_{3 \times 3} & & & & \\ & & & & [Q]_{3 \times 3} & & & \\ & & & & & [Q]_{3 \times 3} & & \\ & & & & & & [Q]_{3 \times 3} & \\ & & & & & & & [Q]_{3 \times 3} \end{bmatrix} \tag{9.75}$$

It is worth noting that this technique can also be used to rotate any planar element including triangle elements.

9.6 Code Implementation

In order to implement the shell element, we need to introduce several changes to the code. First, we update the **defElems** function. Next, we generate shear and bending B-matrices for use with the plate element and a rotation function for use with the shell element. Finally, we code in both the plate and shell elements.

9.6.1 Modifications to the Element Definition Function

We update the **defElems** function by replicating the six lines defining the two new element functions (**Ke_plate** and **Ke_shell**) and updating the active dofs in each dimension.

```
% 6. Plate
net = net + 1;
kList{net} = 'Ke_plate';
iad(net,:,1) = [0 0 0 0 0 0 0]; % 1D
iad(net,:,2) = [0 0 1 1 1 0 0]; % 2D
iad(net,:,3) = [0 0 0 0 0 0 0]; % 3D

% 7. Shell
net = net + 1;
kList{net} = 'Ke_shell';
iad(net,:,1) = [0 0 0 0 0 0 0]; % 1D
iad(net,:,2) = [1 1 1 1 1 1 0]; % 2D
iad(net,:,3) = [1 1 1 1 1 1 0]; % 3D
```

9.6.2 Plate B-matrices

To simplify our code, we generate both the bending and shear B-matrices for the plate element. We generate the two B-matrices individually, starting with the bending B-matrix:

```
1 function BM = B_bending(B)
2
3 nen = size(B,2);
4 for j = 1:nen
5   BM(:,3*j-1:3*j) = [0 -B(1,j); B(2,j) 0; B(1,j) -B(2,j)];
6 end
```

The shear B-matrix uses both the original B-matrix and the shape functions, so we need to pass both variables to the function:

```
1 function BM = B_shear(B,N)
2
3 nen = size(B,2);
4 for j = 1:nen
5   BV(:,3*j-2:3*j) = [B(1,j) 0 N(j); B(2,j) -N(j) 0];
6 end
```

9.6.3 Rotating a Planar Element

We introduce a rotation function to generate both a rotation matrix and local coordinates given a set of element coordinates in the global coordinate system.

```
 1 function [Qe,xe] = rotPlane(xe)
 2
 3 nen = size(xe,1);                              % number of element nodes
 4 nsd = size(xe,2);                         % number of spatial dimensions
 5
 6 for i = 1:nen
 7   xe(i,:) = xe(i,:) - xe(1,:);                   % set first node as origin
 8 end
 9
10 Qe = eye(nsd);
11 if nsd == 3             % only rotate if in three spatial dimensions
12   nx = xe(2,:)/norm(xe(2,:));
13   nz = cross(nx,xe(3,:))/norm(cross(nx,xe(3,:)));
14   ny = cross(nz,nx);
15   Qe = [nx;ny;nz];
16 end
17
18 for i = 1:nen
19   xe(i,:) = xe(i,:)*Qe';        % rotate element into local coordinates
20 end
21 xe = xe(:,1:2);                             % isolate coordinates in XY plane
```

Note that the local coordinates are returned in an assumed 2D coordinate system. If the element supplied is non-planar, the function will only use the first three nodes for alignment; if any of the remaining nodes are out-of-plane, their local z-axes components will be discarded.

9.6.4 Plate Element

The plate stiffness function, `Ke_plate`, very closely parallels the structure of the elastic stiffness function. Once again, we decompose the D-matrix and integration into two components; this time, however, the two components correspond to bending and shear rather than first and second Lamé parameters. It is not necessary to differentiate between element types since we have only developed a quad plate element.

```
 1 function [Ke,D,Te] = Ke_plate(xe,prop)
 2
 3 nen = size(xe,1);                              % number of element nodes
 4 npb = prop(14);                    % number of bending integrations points
 5 nps = prop(15);                     % number of shear integrations points
 6
 7 % Generate constitutive matrix
 8 E = prop(1); v = prop(6); t = prop(7);
 9 DM = E/(1-v^2)*[1 v 0; v 1 0; 0 0 (1-v)/2];       % bending D-matrix
10 DV = E/(2+2*v)*eye(2);                            % shear D-matrix
11 D = blkdiag(DM,DV);                               % complete D-matrix
12
13 % Use gaussian integration to find bending component of ke
14 ke = zeros(nen*3);
15 [ptb,wtb] = genGauss(npb);
16 for i = 1:size(ptb,1)
17    [B,J] = genShape(xe,ptb(i,:));
18    BM = B_bending(B);
19    ke = ke + (t^3)/12*det(J)*BM'*DM*BM*wtb(i);
20 end
21
22 % Use gaussian integration to find shear component of ke
23 [ptv,wtv] = genGauss(nps);
24 for i = 1:size(ptv,1)
25    [B,detJ,N] = genShape(xe,ptv(i,:));
26    BV = B_shear(B,N);
27    ke = ke + t*5/6*detJ*BV'*DV*BV*wtv(i);
28 end
29
30 Te = 1;
31 Ke = Te'*ke*Te;
```

9.6.5 Shell Element

The shell stiffness function, Ke_shell, borrows its framework from MSA and FEM element functions. As with any MSA element, we begin by rotating the shell element into local coordinates. Next, we generate the elastic, kE, and plate, kP, stiffness contributions by calling on the respective element stiffness functions directly. The drilling stiffness is extracted from the diagonals of the elastic and plate elements. We also output the full constitutive matrix, D, consisting of the elastic, DE, and plate, DP, contributions.

```
1 function [Ke,D,Te] = Ke_shell(xe,prop)
2
3 nen = size(xe,1);                                        % number of element nodes
4
5 % Rotate element into local dimensions
6 [Qe,xe] = rotPlane(xe);
7
8 % Generate transformation matrix, Te
9 Te = blkdiag(Qe,Qe,Qe,Qe,Qe,Qe,Qe,Qe);
10
11 % Find stiffness components
12 [kE,DE] = Ke_elastic(xe,prop);                           % membrane component
13 [kP,DP] = Ke_plate(xe,prop);                             % plate component
14 kT = min(abs([diag(kP);diag(kE)]))/1000;                 % twist component
15 D = blkdiag(DE,DP);
16
17 % Assemble elements out of membrane, plate, and twist components
18 ke = zeros(nen*6);
19 for i = 1:4
20   for j = 1:4
21     ke(6*i-5:6*i-4,6*j-5:6*j-4) = kE(2*i-1:2*i,2*j-1:2*j);
22     ke(6*i-3:6*i-1,6*j-3:6*j-1) = kP(3*i-2:3*i,3*j-2:3*j);
23   end
24   ke(6*i,6*i) = kT;
25 end
26 Ke = Te'*ke*Te;
```

9.7   Plate Example

To illustrate the aptitude of the plate element, we analyze one quarter of a simply-supported square plate subject to a downward distributed load, $p = 1\,\text{kPa}$:

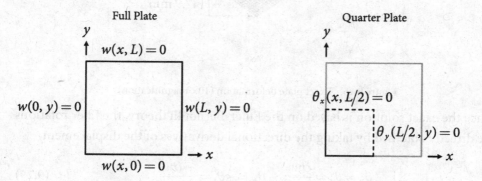

Figure 9.24.   Simply-supported plate geometry.

For our analysis, we use the following geometric and material properties:

$$L = 10\,\text{m}; \quad t = 100\,\text{mm}; \quad E = 30\,\text{GPa}; \quad v = 0.3 \tag{9.76}$$

9.7.1 Exact Solution

The exact solution for this problem was first established by Timoshenko:

$$w = -\frac{16L^4 p}{\pi^6 D} \sum_{m=1:\infty} \sum_{n=1:\infty} \left(\frac{\sin\left(m\pi x/L\right)\sin\left(n\pi y/L\right)}{m^5 n + 2m^3 n^3 + mn^5} \right) \tag{9.77}$$

The variable D is the **flexural rigidity** of the plate defined as follows:

$$D = \frac{Eh^3}{12(1-v^2)} \tag{9.78}$$

This solution converges quickly; after only three iterations, the midspan deflection is accurate to four significant digits. We use 24 iterations to ensure the exact solution is accurate to a high number of significant digits at every point:

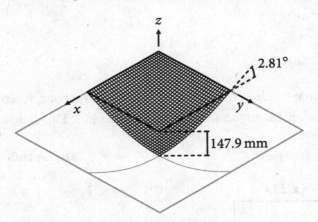

Figure 9.25. Exact plate deformation (10× magnification).

Since the exact solution is based on the Euler-Bernoulli theory, the fiber rotations are directly obtained by taking the directional derivatives of the displacement:

$$\theta_x = \theta_2 = \frac{\partial w}{\partial y}; \quad \theta_y = -\theta_1 = -\frac{\partial w}{\partial x} \tag{9.79}$$

We plot the exact vertical displacements and directional rotations in the xy-plane for the bottom left quarter of the plate.

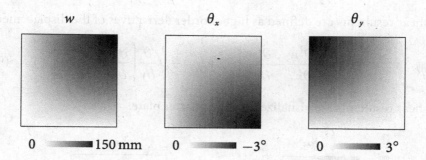

Figure 9.26. Vertical displacements and directional rotations for quarter plate.

It is worth noting that even though the two rotations are opposite in sign, they are symmetrical in magnitude across the xy-diagonal:

$$\theta_x(x, y) = -\theta_y(y, x) \tag{9.80}$$

The moment resultants are defined as higher order derivatives of displacement:

$$m_x = D\left(\frac{\partial^2 w}{\partial x^2} + v\frac{\partial^2 w}{\partial y^2}\right); \quad m_y = D\left(\frac{\partial^2 w}{\partial y^2} + v\frac{\partial^2 w}{\partial x^2}\right) \tag{9.81}$$

$$m_{xy} = D(1-v)\frac{\partial^2 w}{\partial x \partial y} \tag{9.82}$$

Below, we plot the exact moment resultants for the quarter plate:

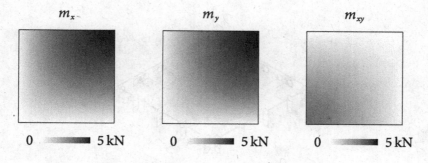

Figure 9.27. Moment resultants for quarter plate.

The first two moment resultants are symmetrical across the xy-diagonal.

$$m_x(x, y) = m_y(y, x) \tag{9.83}$$

The shear resultants are defined as higher order derivatives of the displacement:

$$v_{xz} = -D\frac{\partial}{\partial x}\left(\frac{\partial^2 w}{\partial x^2} + \frac{\partial^2 w}{\partial y^2}\right); \quad v_{yz} = -D\frac{\partial}{\partial y}\left(\frac{\partial^2 w}{\partial x^2} + \frac{\partial^2 w}{\partial y^2}\right) \tag{9.84}$$

The shear resultants are visualized for the quarter plate:

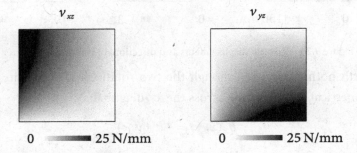

Figure 9.28.   Shear resultants for quarter plate.

The shear resultants are also symmetrical:

$$v_{xz}(x, y) = v_{yz}(y, x) \tag{9.85}$$

9.7.2 FEM Implementation

For FEM implementation, we must first discretize the problem geometry:

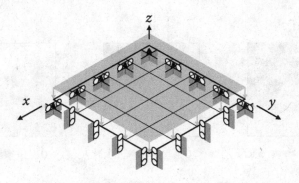

Figure 9.29.   Discretized system.

Along the simply-supported boundaries ($x = 0$ and $y = 0$), we restrain both the vertical deflections and the rotation parallel to associated side:

$$w(0, y) = \theta_x(0, y) = 0; \quad w(x, 0) = \theta_y(x, 0) = 0 \tag{9.86}$$

Along the lines of symmetry ($x = L/2$ and $y = L/2$), we restrain rotations perpendicular to the associated axis of symmetry:

$$\theta_y(L/2, y) = 0; \quad \theta_x(x, L/2) = 0 \tag{9.87}$$

The natural BCs are defined by the applied distributed load. Since all of the elements are squares and the distributed load is uniform, each element experiences a total element load, P_e, equal to the product of the element area ($A^e = L_x^e L_y^e$) and the distributed load. When we sum the contributions at each node, the internal nodes will experience the full element load, the boundary nodes will experience half of the full element load, and the corner nodes will experience only one quarter of the full element load.

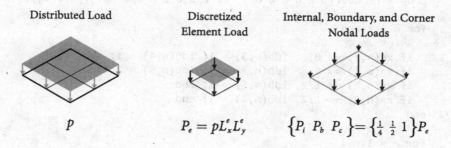

Distributed Load	Discretized Element Load	Internal, Boundary, and Corner Nodal Loads
p	$P_e = pL_x^e L_y^e$	$\left\{ P_i \; P_b \; P_c \right\} = \left\{ \frac{1}{4} \; \frac{1}{2} \; 1 \right\} P_e$

Figure 9.30. Generating nodal loads.

Below, we present the main input script for this example. We use the established mesh generation function (**genMesh**) to produce the discretized mesh. Conditional statements are used to assign essential BCs to appropriate nodes. The natural BCs take the form of concentrated loads summed from the nodal contributions of each element. Although loads are applied to vertically restrained nodes, these loads are ignored because the script in prioritizes essential over natural BCs.

```
 1 % exChpt9Plate - Analysis of a square plate in N and mm base units
 2 clear;
 3
 4 % 0. Convergence inputs
 5 neX = 16;                          % number of elements along x-axis
 6 neY = neX;                         % number of elements along y-axis
 7 npb = 4;                           % number of bending integration pts
 8 npv = 4;                           % number of shear integration pts
 9
10 % 1. Global definitions
11 nsd = 2;                                % number of spatial dimensions
12 ndf = 7;                                % number of degrees of freedom
13 L = 10000; t = 100; E = 30000; v = 0.3;
14 p = -0.01; Pe = p*(L/2)^2/(neX*neY);
15
16 % 2. Nodal & 3. Element definitions
17 [xn,ien] = genMesh(L/2,L/2,neX,neY,nsd,3);
18 nnp = size(xn,1);                        % number of nodal points
19 nel = size(ien,1);                       % number of elements
20 idb = zeros(nnp,ndf);                    % index of dofs - supported
21 ds = zeros(nnp,ndf);       % prescribed displacements at supports (mm)
22 Pu = zeros(nnp,ndf);       % applied forces at unrestrained dofs (N,Nmm)
23 prop = repmat([6 E 0 0 0 0 v t 0 0 0 0 0 0 npb npv],[nel 1]);
24
25 for n = 1:nnp
26    % Set essential BCs
27    if xn(n,1) == 0,   idb(n,3) = 1; idb(n,4) = 1; end
28    if xn(n,2) == 0,   idb(n,3) = 1; idb(n,5) = 1; end
29    if xn(n,1) == L/2, idb(n,5) = 1; end
30    if xn(n,2) == L/2, idb(n,4) = 1; end
31 end
32
33 for e = 1:nel                                     % Set  natural BCs
34    for i = 1:4
35       Pu(ien(e,i),3) = Pu(ien(e,i),3) + Pe/4;
36    end
37 end
38
39 % 4. RUN ANALYSIS
40 [results,process] = runAnalysis(Pu,ds,xn,prop,idb,ien);
41 [F,Rs,Fe,Fi,d,du,de] = deal(results{:});
42 [Kuu,Ke,ke,Te,ied,idu,ids] = deal(process{:});
```

9.7.3 Results

We begin our analysis with a convergence study to evaluate integration schemes for plate elements before plotting the displacements and stress resultants.

9.7.3.1 Convergence Study

Our first task is to choose the best integration scheme for the plate element. Since the plate is thin, the shear stiffness is expected to scale poorly relative to the bending stiffness, thus leading to **shear locking**. To overcome this issue, we use selective reduced integration: under-integrating the shear component but fully integrating the bending component. For comparison, we also show convergence using full 2×2 integration and uniform reduced 1×1 integration schemes. We limit our convergence study to the deformation of the center of the plate and the unrestrained rotation at the midpoint of a side.

nel	Selective (2x2 & 1x1) $\tilde{w}/w$	Selective (2x2 & 1x1) θ/θ	Full (2x2 & 2x2) $\tilde{w}/w$	Full (2x2 & 2x2) θ/θ	Reduced (1x1 & 1x1) $\tilde{w}/w$	Reduced (1x1 & 1x1) θ/θ
1x1	0.786	0.946	0.003	0.002	0.962	1.159
2x2	0.978	0.994	0.011	0.011	1.016	1.030
4x4	0.995	0.998	0.043	0.041	1.004	1.007
8x8	0.999	1.000	0.151	0.147	1.001	1.002
16x16	1.000	1.000	0.416	0.408	1.001	1.000
32x32	1.000	1.000	0.741	0.733	1.001	1.000
64x64	1.000	1.000	0.920	0.917	1.001	1.000

The convergence of the vertical displacement and rotation follows a similar trend, highlighting the consistency of the approximation. We plot the convergence profile for vertical displacement of the plate center for all three integration schemes.

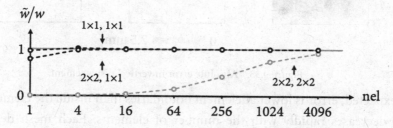

Figure 9.31. Convergence of vertical displacement at plate midspan.

Both selective and uniform reduced integration schemes perform significantly better than the full integration scheme. As with plane strain volumetric locking, uniform reduced integration is unreliable as it can produce zero-energy modes.

9.7.3.2 Displacements

The plate's displacement is best characterized by its deformed geometry. Even though the plate geometry is two-dimensional, the deformation occurs out of plane; hence, the deformed shape is best presented using a three-dimensional, isometric projection. For these displacements and the subsequent analysis, we use three meshes (composed from 4, 16, and 32 elements).

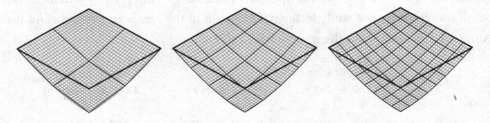

Figure 9.32.   Deformed plate using 2×2, 4×4, and 8×8 meshes (10× magnification).

Since we have the exact solution available to us, we can plot the error in the displacement. Note that although the origin in the isometric projection is located at the top-center of the image, it is located at the bottom-left of the planar plots.

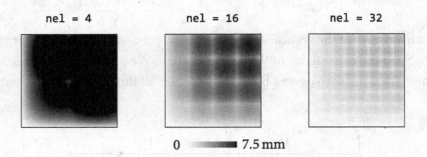

Figure 9.33.   Absolute error in vertical displacement.

As expected, error is lower at element boundaries than inside the element domain and decreases rapidly with the number of elements. Each mesh demonstrates symmetry about the xy-diagonal and larger error towards the plate center.

The theoretical directional rotations are symmetrical in magnitude across the xy-diagonal. Due to the regularity of the FEM grid, this symmetry is reflected in the analysis. Hence, we limit our investigation to the rotation about the x-axis.

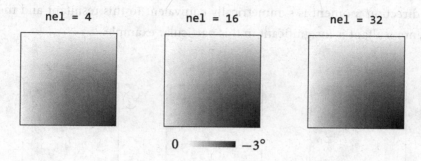

Figure 9.34. Rotation about x-axis.

The rotation is continuous across element boundaries, as expected from a primary unknown. The largest rotations occur along the bottom boundary. In these plots, the difference between the various meshes are difficult to decipher; we turn to the error plots to see how the mesh affects the accuracy of the rotations.

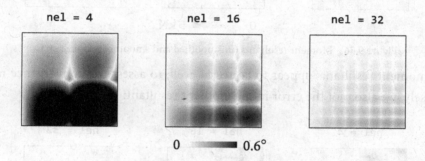

Figure 9.35. Absolute error in rotation about x-axis.

The error in the rotation is smallest at the edges and largest within the element, again characterizing the behavior of a primary unknown.

9.7.3.3 Resultants

We next investigate the moment and shear resultants from our analysis. Since we expect both sets of resultants to be discontinuous, we provide both the unsmoothed and smoothed plots for each.

Though we have three moment resultants, we limit our investigation to the x-direction moment resultant. We ignore the other two moment resultants because the y-direction moment is symmetrically equivalent to this resultant and the xy-moment resultant is insignificant in this particular example.

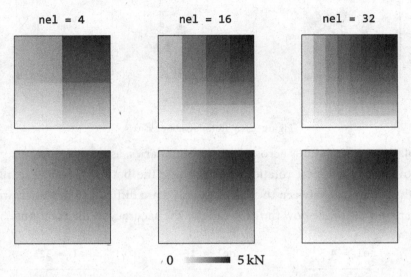

Figure 9.36.   Moment resultants (unsmoothed and smoothed) about x-axis.

The moment resultants appear to converge well; to assess the convergence more precisely, we also plot the error in the moment resultant:

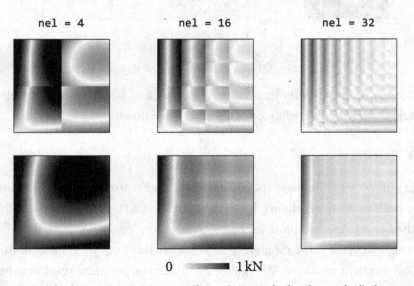

Figure 9.37.   Absolute error in moment resultants (unsmoothed and smoothed) about x-axis.

As expected, the error reduces as the mesh becomes finer. We note that the error in the unsmoothed plots is largest at the element boundaries and smallest within the element domain, as expected from a derivative of a primary variable. Observing the smoothed profiles, we note that only local spikes are reduced, but the overall magnitude of the error is unaffected. While smoothing does improve behavior along element boundaries, it is not particularly reliable at the boundaries of the BVP; for coarser meshes (i.e., nel = 4), the number of internal nodes is significantly smaller than the number of boundary nodes leading to decreased performance in the smoothed mesh.

We next look at shear resultants. Noting the symmetrical equivalence between xz- and yz-direction shear resultants, we only present the xz-resultants.

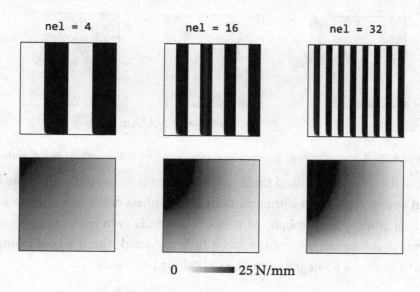

Figure 9.38. Shear resultants (unsmoothed and smoothed) in xz-direction.

These unsmoothed shear resultant plots distinctly demonstrate shear locking in the plate. While we were able to avoid the effects of shear locking by using selective reduced integration, this technique only eliminates the *effect*, not the *presence* of shear locking. Since the magnitude of the displacements using selective reduced integration is significantly larger than the displacements found using a fully-integrated analysis, shear locking reappears in our analysis as greatly exaggerated shear resultants. To repair these shear resultants, we need to use a smoothing technique consistent with the integration scheme. Hence, we use 1×1 rather than 2×2 integration for smoothing to obtain the much improved smoothed shear resultant plots presented above. To demonstrate shear locking and the improvement due to smoothing, we plot the shear resultant error.

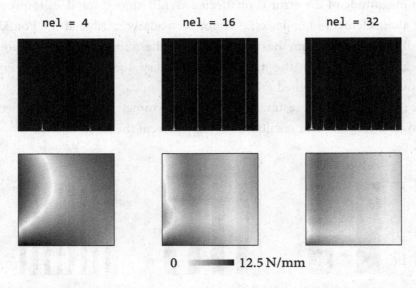

Figure 9.39.   Absolute error in shear resultants (unsmoothed and smoothed) in x-direction.

We note that the unsmoothed shear resultant error is enormous and off the error scale at nearly every point within the domain regardless of the mesh fineness. Only vertical white lines through the element centroids (where the single point of integration is located) appear not to be fully saturated, thereby confirming the choice of using 1×1 integration in the smoothing procedure.

9.8   Shell Example

Having demonstrated the efficacy of the plate element, we next investigate the performance of the shell element. Since we are dealing with more complex elements, theoretical solutions are increasingly rare. For our analysis we will use the standard benchmark test for shells called the **Scordelis-Lo roof problem**, which describes the behavior of a cylindrical roof that is simply-supported along the end arches and subject to a uniform gravity load.

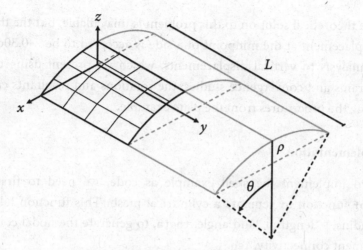

Figure 9.40. Scordelis-Lo roof problem.

The standard dimensions of this problem are presented using imperial units:

$$L = 50 \text{ ft} ; \quad \rho = 25 \text{ ft} ; \quad t = 0.25 \text{ ft} ; \quad \theta = 40° \tag{9.88}$$

The material properties are provided as follows:

$$E = 432 \times 10^6 \text{ psf} ; \quad v = 0.0 \tag{9.89}$$

The natural BCs of the problem are limited to a downward distributed load applied uniformly over the area of the roof:

$$p = 90 \text{ psf} \tag{9.90}$$

The essential BCs restrain the two side arches from translating in vertical and longitudinal directions and rotating about the y-axis:

$$u(x,0) = w(x,0) = \theta_y(x,0) = 0; \quad u(x,L) = w(x,L) = \theta_y(x,L) = 0 \tag{9.91}$$

We reduce the computational burden by taking advantage of symmetry to model a quarter of the shell. First, we capture symmetry along the y-axis, $x = 0$:

$$u(0,y) = \theta_y(0,y) = \theta_z(0,y) = 0 \tag{9.92}$$

We introduce another symmetry restraint along the midsection cut, $y = L/2$:

$$u(x,L/2) = \theta_x(x,L/2) = \theta_z(x,L/2) = 0 \tag{9.93}$$

A complete theoretical solution to this problem is unavailable, but the theoretical vertical displacement at the midpoint of a side is accepted to be $-0.3068\,\text{ft}$. We limit our analysis to vertical displacements, which we present using deformed shape diagrams and convergence studies; the rotations and resultants can be extracted using the procedures from the plate example.

9.8.1 Implementation

In order to implement the shell example as code, we need to first write a **genMeshRoof** function to generate a cylindrical mesh. This function takes as inputs the radius, R, length, L, and angle, theta, to generate the nodal coordinates, xn, and element connectivity, ien.

```
 1 function [xn,ien] = genMeshRoof(R,L,theta,neR,neY)
 2
 3 nel = neR*neY;                                          % number of elements
 4 nen = 4;                                          % number of element nodes
 5 nnp = (neR+1)*(neY+1);                              % number of nodal points
 6
 7 % nodal definitions
 8 Tinc = theta/neR;
 9 Yinc = L/neY;
10 xn = zeros(nnp,3);
11 for i = 1:neR+1
12   for j = 1:neY+1
13     n = i+(j-1)*(neR+1);
14     thetaN = Tinc*(i-1);
15     xn(n,:) = [R*sin(thetaN) (j-1)*Yinc R*cos(thetaN)];
16   end
17 end
18
19 % element definitions
20 ien = zeros(nel,nen);                               % index of element nodes
21 for i = 1:neR
22   for j = 1:neY
23     e = i+(j-1)*neR;
24     n1 = i+(j-1)*(neR+1); n2 = n1 + 1;
25     n3 = i+j*(neR+1); n4 = n3 + 1;
26     ien(e,:)     = [n1 n2 n4 n3];
27   end
28 end
```

Next, we update the main script used for the plate example with the new material properties, the new meshing algorithm, **genMeshRoof**, and the updated element load, **Pe**.

```
1  % exChpt9Shell - Analysis of Scordelis-Lo roof in ft & lb base units
2  clear;
3
4  % 0. Convergence inputs
5  neR = 1;                              % number of elements along x-axis
6  neY = neR;                            % number of elements along y-axis
7  npl = 4; npm = 4;                % membrane (lambda, mu) integration pts
8  npb = 4; npv = 1;               % plate (bending, shear) integration pts
9
10 % 1. Global definitions
11 nsd = 3;                              % number of spatial dimensions
12 ndf = 7;                              % number of degrees of freedom
13 L = 50; R = 25; t = 0.25; E = 4.32e8; v = 0; theta = 40*(pi/180);
14 q = -90; Pe = q*(L/neY/2)*(2*R*sin(1*pi/9/neR));
15
16 % 2. Nodal & 3. Element definitions
17 [xn,ien] = genMeshRoof(R,L/2,theta,neR,neY);
18 nnp = size(xn,1);                         % number of nodal points
19 nel = size(ien,1);                          % number of elements
20 idb = zeros(nnp,ndf);                 % index of dofs - supported
21 ds = zeros(nnp,ndf);
22 Pu = zeros(nnp,ndf);
23 prop = repmat([7 E 0 0 0 0 v t 0 0 0 0 npl npm npb npv],[nel 1]);
24
25 for n = 1:nnp                                   % Set essential BCs
26    if xn(n,2) == 0,   idb(n,[1 3 5]) = 1; end
27    if xn(n,1) == 0,   idb(n,[1 5 6]) = 1; end
28    if xn(n,2) == L/2, idb(n,[2 4 6]) = 1; end
29 end
30
31 for e = 1:nel                                    % Set   natural BCs
32    for i = 1:4
33      Pu(ien(e,i),3) = Pu(ien(e,i),3) + Pe/4;
34    end
35 end
36
37 % 4. RUN ANALYSIS
38 [results,process] = runAnalysis(Pu,ds,xn,prop,idb,ien);
39 [F,Rs,Fe,Fi,d,du,de] = deal(results{:});
40 [Kuu,Ke,ke,Te,ied,idu,ids] = deal(process{:});;
```

9.8.2 Results

We limit our investigation to a convergence study and plots of the deformed shape. We begin with a convergence study of the theoretical vertical displacement at the midpoint of a side:

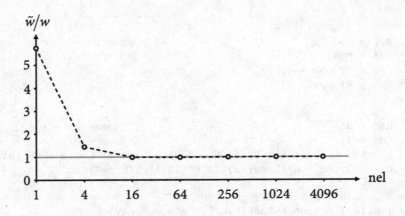

Figure 9.41.   Scordelis-Lo roof convergence.

Since the roof is described by an extruded smooth curve, no discretized mesh will exactly replicate this continuous geometry. As a result, refining the mesh induces convergence in both the geometry and the modelled behavior. Consequently, coarse meshes perform quite poorly, but convergence progresses quickly.

To evaluate the performance of the various meshes, we use the 32×32 mesh (nel = 1024) as a reliable approximation of the exact solution:

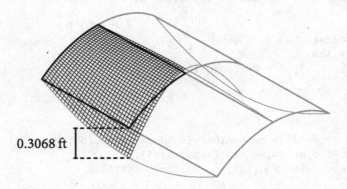

Figure 9.42.   Deformed geometry using 1024 element mesh (25× magnification).

At the roof midspan ($y = 12.5\,\text{ft}$), the edge displaces downward while the arch crown moves upward. This behavior is indicative of the more complex properties of shell structures. In this particular example, the significant deformation of the unsupported edge leads to a local upward pinching effect through the shell cross-section, which is ultimately more significant than the global downward deflection due to the gravity load. Superimposing the deformed shape for three iterations (4, 16, and 64) we observe how quickly the approximation converges:

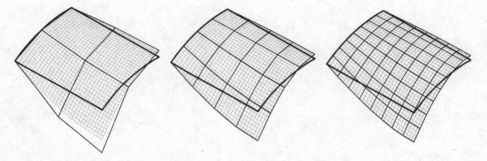

Figure 9.43. Deformed geometry using 4, 16, and 64 element meshes (25× magnification).

Appendix A

Code Supplement

In this appendix, we present complete formulations for the element definition and internal force functions. We have already developed all of the components in the element definition function, `defElems`, so we only need summarize it in this appendix. Although we introduced the `getIntern` function in Chapter 4, we did not update it so we present it completely for the first time in this appendix.

A.1 Element Definition Function

In this text, we developed a total of seven elements: three in MSA (truss, beam, and frame) and four in FEM (heat, elastic, plate, and shell). We present the complete `defElems` function below:

```
 1 function [kList,iad] = defElems
 2
 3 net = 1;                                           % number of element types
 4 iad = zeros(net,ndf,3);        % index of activated degrees of freedom
 5 typ = 0;                             % counter to track element type
 6
 7 % 1. Truss
 8 typ = typ + 1;
 9 kList{typ} = 'Ke_truss';
10 iad(typ,:,1) = [1 0 0 0 0 0 0];                                  % 1D
11 iad(typ,:,2) = [1 1 0 0 0 0 0];                                  % 2D
12 iad(typ,:,3) = [1 1 1 0 0 0 0];                                  % 3D
13
14 % 2. Beam
15 typ = typ + 1;
16 kList{typ} = 'Ke_beam';
17 iad(typ,:,1) = [0 1 0 0 0 1 0];                                  % 1D
18 iad(typ,:,2) = [0 0 1 1 1 0 0];                                  % 2D
19 iad(typ,:,3) = [0 0 0 0 0 0 0];                                  % 3D
```

```
20 % 3. Frame
21 typ = typ + 1;
22 kList{typ} = 'Ke_frame';
23 iad(typ,:,1) = [1 1 0 0 0 1 0];          % 1D
24 iad(typ,:,2) = [1 1 0 0 0 1 0];          % 2D
25 iad(typ,:,3) = [1 1 1 1 1 1 0];          % 3D
26
27 % 4. Heat
28 typ = typ + 1;
29 kList{typ} = 'Ke_heat';
30 iad(typ,:,1) = [0 0 0 0 0 0 1];          % 1D
31 iad(typ,:,2) = [0 0 0 0 0 0 1];          % 2D
32 iad(typ,:,3) = [0 0 0 0 0 0 0];          % 3D
33
34 % 5. Membrane
35 typ = typ + 1;
36 kList{typ} = 'Ke_elastic';
37 iad(typ,:,1) = [0 0 0 0 0 0 0];          % 1D
38 iad(typ,:,2) = [1 1 0 0 0 0 0];          % 2D
39 iad(typ,:,3) = [0 0 0 0 0 0 0];          % 3D
40
41 % 6. Plate
42 typ = typ + 1;
43 kList{typ} = 'Ke_plate';
44 iad(typ,:,1) = [0 0 0 0 0 0 0];          % 1D
45 iad(typ,:,2) = [0 0 1 1 1 0 0];          % 2D
46 iad(typ,:,3) = [0 0 0 0 0 0 0];          % 3D
47
48 % 7. Shell
49 typ = typ + 1;
50 kList{typ} = 'Ke_shell';
51 iad(typ,:,1) = [0 0 0 0 0 0 0];          % 1D
52 iad(typ,:,2) = [1 1 1 1 1 1 0];          % 2D
53 iad(typ,:,3) = [1 1 1 1 1 1 0];          % 3D
```

A.2 Internal Forces

Because the principal objective of MSA and FEM analysis is obtaining unknown displacements and reactions, we need to perform post-processing to extract further information about the system's structural behavior. We set up the **getIntern** function to extract internal forces, stresses, and resultants.

```
1 function [Fi] = getIntern(Fe,de,ke,Te,xn,prop,ien)
2
3 nsd = size(xn,2);                % number of spatial dimensions
4 nel = size(prop,1);              % number of elements
```

```
 5 nen = sum(ien > 0,2);                          % number of element nodes
 6 [F,Vy,Mz,Vz,My,T,sx,sy,sxy,vyz,vxz,mx,my,mxy] = deal(zeros(nel,1));
 7
 8 for e = 1:nel
 9   typ = prop(e,1); E = prop(e,2); A = prop(e,3); t = prop(e,8);
10   xe = xn(ien(e,1:nen(e)),:);              % element nodal coordinates
11   di = Te{e}*de{e};              % displacements in local coordinates
12   f = Te{e}*Fe{e};                          % forces in local coordinates
13   switch typ
14     case 1                                                    % Truss
15       F(e) = -f(1);
16       sx(e) = F(e)/(A*E);
17     case 2                                                    % Beam
18       if nsd == 1
19         Vy(e) = f(1);
20         Mz(e,1:2) = [-f(2) f(4)];
21       else
22         Vz(e) = f(1);
23         My(e,1:2) = [-f(3) f(6)];
24         T(e) = -f(2);
25       end
26     case 3                                                    % Frame
27       if nsd < 3
28         F(e) = -f(1);
29         Vy(e) = f(2);
30         Mz(e,1:2) = [-f(3) f(6)];
31       else
32         F(e) = -f(1);
33         Vy(e) = f(3);
34         Mz(e,1:2) = [-f(5) f(11)];
35         Vz(e) = -f(4);
36         My(e,1:2) = [f(2) -f(6)];
37         T(e) = f(12);
38       end
39   end
40
41   if typ > 4          % strains/stresses for membrane, plate, and shell
42     switch nen(e)
43       case 2, pts = [-1; 1];
44       case 3, pts = [-1 -1;1 -1;1 1];
45       case 4, pts = [-1 -1;1 -1;1 1;-1 1];
46     end
47     [~,xe] = rotPlane(xe);
48     D = ke{e};
49     for i = 1:nen(e)
50       [B,~,N] = genShape(xe,pts(i,:));
51       switch typ
52         case 5                                                % Membrane
53           BE = B_elastic(B);
```

```
54              epsE = BE*di;
55              sigE = D*epsE;
56              % Populate outputs
57              sx(e,i) = sigE(1); sy(e,i) = sigE(2); sxy(e,i) = sigE(3);
58          case 6                                        % Plate
59              BP = [B_bending(B); B_shear(B,N)];
60              epsP = BP*di;
61              sigP = D*[t*epsP(1:2); -t^3/12*epsP(3:5)];
62              % Populate outputs
63              vyz(e,i)= sigP(1); vxz(e,i)= sigP(2);
64              mx(e,i) = sigP(3); my(e,i) = sigP(4); mxy(e,i) = sigP(5);
65          case 7                                        % Shell
66              BE = B_elastic(B);
67              BP = [B_bending(B); B_shear(B,N)];
68              de = di([1 2 7 8 13 14 19 20]);        % membrane disp
69              dp = di([3 5 6 9 11 12 15 17 18 21 23 24]);  % plate disp
70              epsS = [BE*de; BP*dp];
71              sigS = D*[epsS(1:3); t*epsS(4:6); -t^3/12*epsS(7:8)];
72              % Populate outputs
73              sx(e,i) = sigS(1); sy(e,i) = sigS(2); sxy(e,i) = sigS(3);
74              vyz(e,i)= sigS(4); vxz(e,i)= sigS(5);
75              mx(e,i) = sigS(6); my(e,i) = sigS(7); mxy(e,i) = sigS(8);
76          end
77      end
78    end
79 end
80 Fi = cell(14,1);
81 [Fi{:}] = deal(F,Vy,Mz,Vz,My,T,sx,sy,sxy,vyz,vxz,mx,my,mxy);
```

With the exception of the truss element, we do not extract any stresses for MSA elements. Conversely, we do not extract any forces for FEM elements. For elastic triangle and quad elements, we extract the three directional stresses at each of the element nodes. For the plate element, we extract the two shear and three moment resultants at each element node. For the shell element, we extract both stresses and stress resultants as we did for the membrane and plate elements respectively. We do not generate any internal forces or fluxes for the heat element.

The internal forces, stresses, and resultants we generate with this function supply all of the values used for the examples covered in this text; even the stresses and resultants are provided at nodes to permit smoothing with the **smtMesh** functions. This function is flexible and can be readily updated for a particular analysis.

Appendix B

Nomenclature

This appendix catalogues the symbols and variables used throughout the text. In summarizing the book's terminology, this appendix also serves as a summary of the book content itself. Although the symbols in the text translate consistently to the syntax in the code, there are certain variables that remain exclusive to one set or the other; hence, we present each set in its entirety.

B.1 Symbols

The symbols used throughout the text can be grouped into several categories: 1) sub/superscripts, 2) matrices, 3) vectors, 4) scalar variables, and 5) diacritics.

B.1.1 Subscripts and Superscripts

We begin with symbols that appear consistently as sub/superscripts that can be generally categorized as indices and identifiers. We distinguish that **indices** enumerate instances of a variable within a list, while **identifiers** define the type or category of a variable.

e	Index (superscript) enumerating elements.
G	Identifier (superscript) for global matrix or vectors.
i, j	Indices (subscripts) typically used to enumerate dofs. These indices identify the row (i) and column (j) of each entry in a stiffness matrix or the entries in a displacement (i) or force (j) vector. Furthermore, these indices are also used to identify a variety of parameters like directional stresses/strains and weight/shape functions.
n, m	Indices (subscripts) enumerating nodes.

p,q	Identifiers (subscripts) for local (element) dofs.
P,Q	Identifiers (subscripts) for global dofs.
s,S	Identifier (subscript) referencing local (s) or global (S) restrained dofs associated with reactions and prescribed displacements.
u,U	Identifier (subscript) referencing local (u) or global (U) active dofs associated with applied loads and unknown displacements.
x,y,z	Identifier (subscript) referencing associated axis or direction.
Γ	Identifier (subscript) referencing dofs lying on BVP boundaries.
Ω	Identifier (subscript) referencing dofs lying in BVP domains.

B.1.2 Matrices

Next, we list variables that appear in matrix form. This list is general with important variations of each matrix classification highlighted as necessary.

$[B]$	B-matrix storing partial derivatives of shape functions. Variations include the elastic B-matrix for membrane elements, $[B^E]$, moment B-matrix for plate elements, $[B^M]$, and the shear B-matrix for plate elements, $[B^V]$.
$[D]$	Constitutive matrix. Variations include elastic plane stress, $[D^\sigma]$, elastic plane strain, $[D^\varepsilon]$, plate moment, $[D^M]$, and plate shear, $[D^V]$, versions.
$[H]$	Smoothing matrix relating discontinuous and smoothed stresses.
$[I]$	Identity matrix.
$[J]$	Jacobian. Square matrix defining the relationship between two coordinate systems. We distinguish the multi-dimensional determinant from its scalar determinant, $|J|$.
$[K]$	Stiffness matrix with entries K_{ij}. Variations include the global stiffness matrix, $[K^G]$, the active stiffness matrix, $[K_{UU}]$, and the element stiffness matrix in both the global, $[K^e]$, and local, $[k^e]$, coordinate systems.
$[\partial\bar{N}]$	Local shape derivative matrix found by taking the partial derivatives of the shape functions with respect to each dimension.
$[Q]$	Rotation matrix relating local and global coordinate systems.

$[S]$ Partial derivative matrix. Variations include the elastic, $[S^E]$, plate bending, $[S^M]$, and plate shear, $[S^V]$, versions.

$[T^e]$ Element transformation matrix relating local and global forces/displacements.

$[x^e]$ Element coordinate matrix storing the coordinates of each element node.

B.1.3 Vectors

Naturally, matrices are followed by vectors. Once again, we identify important variations of each vector as necessary.

$\{d\}$ Displacement vector with entries d_i. Variations include the prescribed displacement vector, $\{d_s\}$, unknown displacement vector, $\{d_U\}$, and element displacement vector both in the global, $\{d^e\}$, and local, $\{\overline{d}^e\}$, coordinate systems.

$\{F\}$ Force vector with entries F_j. Variations include the global force vector, $\{F^G\}$, and the element force vector in both the global, $\{F^e\}$, and local, $\{\overline{F}^e\}$, coordinate systems.

$\{m\}$ Moment resultant vector acting on plates and shells.

$\{n\}$ Directional cosine or surface normal.

$\{P_U\}$ Applied force vector.

$\{q\}$ Heat flux vector. Along natural boundaries, a macron accent designates an applied heat flux vector, $\{\overline{q}\}$.

$\{r\}$ Stress resultants vector. Along natural boundaries, a macron accent designates an applied stress resultant vector, $\{\overline{r}\}$.

$\{R_s\}$ Reaction force vector.

$\{u\}$ Alternate designation for the displacement vector.

$\{v\}$ Shear resultant vector used in plate and shells.

$\{\varepsilon\}$ Strain vector storing normal and shear strains at a single point.

$\{\sigma\}$ Stress vector typically storing normal and shear stresses at a single point. For stress-smoothing, we also define stress vectors which store the discontinuous, $\{\tilde{\sigma}\}$, and smoothed, $\{\breve{\sigma}\}$, stresses for all of the nodes. Along natural boundaries, an applied stress vector is designated, $\{\overline{\sigma}\}$.

B.1.4 Scalar Variables

Next, we take a look at the scalar variables used throughout the text. While we have tried to maintain exclusive nomenclature, several terms share the same variable; these instances are identified below.

A	Area; either the cross-sectional area of a truss or frame element, or the surface area of planar elements.
C	Either a general constant, C_i, or the level of continuity, C^i.
d	General degree of freedom referring to nodal displacements, rotations, or temperatures.
E	Elastic or Young's modulus.
f	Either an internal body forces or a general function.
F	General force with directional variants F_x, F_y, and F_z. It is also used to designate axial force in a truss or frame element.
G	Shear modulus.
h	Height; typically used to define geometry in an example.
I	Moment of inertia used in beams and frames. For 2D beams and 3D frames, we also have directional variants I_y and I_z.
J	Torsional constant used for 2D beams and 3D frames.
k	A general term for stiffness, specifically hinge stiffness, k_h, or thermal conductivity, k.
L	Length; typically used for truss, beam, and frame elements. For triangle and quad elements, side lengths are designated as L_S.
m	Moment resultant with directional variants m_x, m_y, and m_{xy}.
M	Moment; directional variations include M_x, M_y, and M_z.
N_i	Shape function corresponding to dof i.
p	Applied load distributed over length or area.
P	Concentrated applied load.
q	Heat flux.
Q	Volumetric heat source.
$R(x)$	Residual as used in Method of Weighted Residuals.
t	Thickness; typically defining out-of-plane thickness in triangle and quad element.
T	Temperature.
u, v, w	Directional displacements in x-, y-, and z-directions respectively.

U	Either the internal elastic energy or the internal heat conduction term in the heat potential.
v	Shear resultant with directional variants v_{xz} and v_{yz}.
V	Either geometric volume or shear force in beams and frames.
w	Width; typically used to define geometry in an example.
W	Work performed on a system, specifically work performed on the domain, W_Ω, and at the boundary, W_Γ. It also identifies the internal heat source, W_Ω, and external flux, W_Γ, terms in the heat potential.
x, y, z	Cartesian axes or coordinates.
γ	Engineering shear strain with directional variations γ_{xy}, γ_{yz}, and γ_{xz}.
Δ	Delta; a small change typically in length.
δ	Dirac delta.
ε	Strain, particularly normal strains, ε_x, ε_y, and ε_z. It is also used to designate pure shear strains, ε_{xy}, ε_{yz}, and ε_{xz}, but we typically use engineering shear strains instead in this text. Furthermore, principal strains are defined as ε_1, ε_2, and ε_3.
θ	Angle or rotation for beams, frames, plates, and shells with directional variants, θ_x, θ_y, and θ_z. It also designates the principal angle for stress or strain found with Mohr's circle θ_p.
κ	Curvature with directional variants κ_x, κ_y, and κ_{xy}.
λ	First Lamé parameter.
μ	Second Lamé parameter.
ν	Poisson's ratio.
ξ, η, ζ	Parent domain axes.
Π	Total potential; for elasticity it defines total potential energy.
ρ	Radius of curvature.
σ	Stress; particularly normal stresses, σ_x, σ_y, and σ_z. It also designates principal stresses, σ_1, σ_2, and σ_3.
τ	Shear stresses with directional variants τ_x, τ_y, and τ_z.
φ	Angle of inclination for truss, beam, and frame elements. It is also used to designate the shear correction factor φ_V.
Φ	Error functional used in stress smoothing.
$\omega(x)$	Weighting function.
∇	Gradient.

B.1.5 Diacritical Marks

We conclude the list of symbols with several diacritics.

$\breve{\sigma}$	Breve designating the smoothed profile in stress smoothing.
$\hat{n}$	Circumflex designating a normalized vector, typically storing directional cosines or surface normals.
$\bar{x}$	Macron designating localized coordinates. It is also be used to designate heat fluxes, $\bar{q}$, stresses, $\bar{\sigma}$, and stress resultants, $\bar{r}$, applied along BCs.
$\tilde{u}$	Tilde designating an approximate solution to a BVP.

B.2 Code

Having presented the symbols used throughout the text, we continue with the syntax used throughout the code. In presenting code terminology, we have a new set of categories: 1) functions, 2) scripts, 3) counters, 4) numbers, 5) indices, and 6) general variables.

B.2.1 Functions

Our code is structured using functions, which are the natural starting point for presenting the nomenclature in the code. We distinguish between primary functions, which are required for the core analysis, and secondary functions, which are used either for generating inputs or for secondary post-processing. In the following list, secondary functions are halftone.

addForce	Assembles element vectors into a global force vector.
addIndex	Populates the global index based on element connectivity and active element dofs.
addStiff	Assembles the active stiffness matrix from element stiffness matrices.
B_bending	Generates the plate bending B-matrix using the B-matrix.
B_elastic	Generates the elastic B-matrix using the B-matrix.
B_shear	Generates the plate shear B-matrix using the B-matrix and shape functions.
chgShape	Reformats an input force or displacement vector into either a true vector or a matrix indexed corresponding to nodes and dofs.
cntIndex	Enumerates the nonzero entries of an index matrix.

defElems	Defines element function names and their active dofs per spatial dimension.
genForce	Generates an element force vector based on an element displacement vector and the element stiffness matrix.
genGauss	Generates integration points and weights for Gaussian quadrature.
genMesh	Generates a rectangular mesh of specified side lengths using left-biased triangles, right-biased triangles, or quads.
genMeshRoof	Generates a cylindrical mesh of specified radius, angle, and length with left-biased triangles, right-biased triangles, or quads.
genShape	Generates the set of shape function values and their derivatives as well as the Jacobian, B-matrix, and Jacobian determinant for a specific point within an element.
genStress	Generates the strain and stress for a specified point within an element subject to provided displacements.
getIntern	Extracts the internal forces in each element given a global displacement set.
intForce	Integrates a quadratic traction or flux distribution acting along a side of a triangle or quad element.
Ke_beam	Calculates the stiffness of a 1D or 2D two-noded beam element. Conditional statement also incorporates a zero-length hinge element.
Ke_elastic	Calculates the stiffness of a triangle or quad elastic membrane element using a plane stress or plane strain assumption.
Ke_frame	Calculates the stiffness of a 2D or 3D two-noded frame element with the provision for a zero-length hinge element.
Ke_heat	Calculates the stiffness of a rod, triangle, or quad heat element.
Ke_plate	Calculates the stiffness of a quad plate element using the Reissner-Mindlin plate theory.
Ke_shell	Calculates the stiffness of a quad shell element using the Reissner-Mindlin plate theory.
Ke_truss	Calculates the stiffness of a 1D, 2D, or 3D two-noded truss element.
mapParent	Calculates the parent domain coordinates for a point within the physical domain of a given element.
rotPlane	Rotates a planar element (triangle or quad) from the element's local xy-plane.
runAnalysis	Primary script to run any MSA or FEM analysis.
sigPrinc	Calculates the principal stress or strain in an elastic medium as well as the corresponding principal angle.
smtMesh	Smoothes nodal discontinuities (particularly stress, flux, or resultants) in a mesh.

B.2.2 Scripts

All of the examples presented in the text are set up using one of the following scripts, which are named with reference to the chapter in which they appear.

exChpt2	Analysis of a 2D truss in N and mm base units.
exChpt4	Analysis of a 2D truss in N and mm base units.
exChpt5	Analysis of a 2D braced frame in N and mm base units.
exChpt7	Analysis of a 2D heat problem in degC and mm base units.
exChpt8	Analysis of a 2D deep cantilever in N and mm base units.
exChpt9Plate	Analysis of a square plate in N and mm base units.
exChpt9Shell	Analysis of Scordellis-Lo roof in ft and lb base units.

B.2.3 Counters

As with functions, we also have primary and secondary variables: primary variables are globally used and shared by multiple functions; secondary variables are internal to specific functions. In presenting these variables, we also identify the type (integer, Boolean, string, or float) as well as the size. We begin our variable list with counters, which are equivalent to the indices used in the text.

cnt	general counter	integer	[1]
e	element index counter	integer	[1]
i,j	generic counter, specifically for dofs	integer	[1]
m,n	node index counters	integer	[1]
p,q	local dofs index counters	integer	[1]
P,Q	global dofs index counters	integer	[1]
typ	counter for element types	integer	[1]

B.2.4 Numbers

Numbers are used to terminate iterative procedures and typically correspond to the quantity of nodes, elements, points, or dofs.

ned	number of element dimensions	integer	[nel]
nel	number of elements	integer	[1]
nen	number of element nodes	integer	[nel]
neq	number of equations	integer	[1]
neR	number of elements about mesh radius	integer	[1]
net	number of element types	integer	[1]
neX	number of elements on x-axis of mesh	integer	[1]
neY	number of elements on y-axis of mesh	integer	[1]
ndf	number of degrees of freedom	integer	[1]

nip	number of integration points	integer	[1]
nnp	number of nodal points	integer	[nel]
nph	number of integration pts – heat	integer	[1]
npb	number of integration pts – bending	integer	[1]
npl	number of integration pts – lambda	integer	[1]
npu	number of integration pts – mu	integer	[1]
npv	number of integration pts – shear	integer	[1]
nsd	number of space dimensions	integer	[1]

B.2.5 Indices

Indices are used primarily to provide instruction to the code regarding which nodes or dofs are to be included in iterative procedures.

iad	index of active dofs	boolean	[net,ndf,nsd]
ied	index of element dimensions	integer	[nel,ndf]
ien	index of element nodes	boolean	[nel,nsd]
id	general index	int/bool	[nnp,ndf]
idnew	general updated index	int	[nnp,ndf]
idb	index of dofs – restrained	boolean	[nnp,ndf]
ids	index of dofs – supports	integer	[nnp,ndf]
idt	index of dofs – total	boolean	[nnp,ndf]
idu	index of dofs – unknown	integer	[nnp,ndf]

B.2.6 General Variables

As with the text symbols, we have a large collection of code variables. Unlike symbols, there is no typological distinction (such as parenthesis) between scalar and matrix variables. Hence, both types are not distinguished save by their size.

A	cross-sectional area	float	[1]
B	B-matrix	float	[nsd,nen]
BE	B-matrix: elastic	float	[3,2*nen]
BM	B-matrix: plate bending	float	[3,3*nen]
BP	B-matrix: plate total	float	[3,3*nen]
BV	B-matrix: plate shear	float	[2,3*nen]
byRow	reset numbering for each row of index	boolean	[1]
c	coefficients in force integration	float	[3]
center	center of mohr's circle	float	[1]
d	complete displacement vector	float	[nnp,ndf]
dscE	element discontinuous contribution	float	[nen]
dscP	int points discontinuous contribution	float	[nip]
dscNode	discontinuous node	float	[nnp]
de	element displacement vector	float	[nen*ned]
de	membrane component of shell disp	float	[2*nen]

dp	plate component of shell disp	float	[3*nen]
detJ	determinant of Jacobian matrix	float	[1]
di	displacements in local coordinates	float	varies
dN	partial derivatives of shape functions	float	[nsd,nen]
ds	prescribed displacements	float	[nnp,ndf]
du	unknown displacement vector	float	[neq]
D	D-matrix	float	square
DE	D-matrix: membrane component of shell	float	[3,3]
Dl	D-matrix: lambda contribution	float	[3,3]
DM	D-matrix: plate bending contribution	float	[3,3]
DP	D-matrix: plate component of shell	float	[5,5]
Du	D-matrix: mu contribution	float	[3,3]
DV	D-matrix: plate shear contribution	float	[3,3]
eps	strain vector	float	[3]
epsE	elastic strain vector	float	[3]
epsP	plate strain vector	float	[5]
epsS	shell strain vector	float	[8]
expand	change shape to vector or matrix	boolean	[1]
E	Young's modulus	float	[1]
F	force vector, typically active	float	[nnp,ndf]
	also internal axial force	float	[1]
Fe	element force vector	float	[nen*ned]
Fi	internal element forces	float	cell
Fd	force or displacement vector	float	[neq]
Fdnew	reshaped force or displacement vector	float	varies
G	shear modulus	float	[1]
H	global smoothing matrix	float	[nnp,nnp]
He	element smoothing matrix	float	[nnp,nnp]
I	moment of inertia	float	[1]
Iy	moment of inertia about y-axis	float	[1]
Iz	moment of inertia about z-axis	float	[1]
J	Jacobian matrix	float	[nsd,nsd]
	also torsional constant	float	[1]
K	general stiffness matrix	float	square
ke	local element stiffness matrices	float	cell
kE	membrane component of shell stiffness	float	[2*nen,2*nen]
Ke	global element stiffness	float	cell
kP	plate component of shell stiffness	float	[3*nen,3*nen]
kT	torsion component of shell stiffness	float	[nen,nen]
kList	list of function names	string	[net]
kh	hinge stiffness	flaot	[1]
Kuu	active stiffness matrix	float	[neq,neq]
Kus_ds	prescribed displacement contribution	float	[neq]
L	length of linear element	float	[1]
Ls	length of element side	float	[1]
Lx	length of mesh along x-axis	float	[1]
Ly	length of mesh along y-axis	float	[1]
mesh	mesh type (tri-left, tri-right, quad)	integer	[1]
mx,y,xy	moment resultants	float	[1]

My,Mz	moments components in beam or frame	float	[1]
n1...n4	element nodes temporary name	integer	[1]
nx	orientation vector along local x-axis	float	[nsd]
ny	orientation vector along local y-axis	float	[nsd]
nz	orientation vector along local z-axis	float	[nsd]
N	shape functions	float	[nen]
prop	element properties	various	[nel,16]
ptb	integration pts: plate bending	float	[npb,nsd]
ptl	integration pts: elastic lambda	float	[npl,nsd]
pts	points, general or integration	float	[nip,nsd]
ptu	integration pts: elastic mu	float	[npu,nsd]
ptv	integration pts: plate shear	float	[npv,nsd]
PS	plane stress or strain designator	boolean	[1]
Pu	applied force	float	[nnp,ndf]
Qe	element rotation matrix	float	[nsd,nsd]
radius	radius of Mohr's circle	flaot	[1]
Rs	reaction force	float	[nnp,ndf]
s	side ordinates of integration	float	[2]
s1,s2	side ordinates of integration	float	[1]
sig	stress vector	float	[3]
sigE	elastic stress vector	float	[3]
sigP	plate stress vector	float	[5]
	also principal stress vector	float	[2]
sigS	shell stress vector	float	[8]
smtNode	smoothed node stresses	float	[nnp]
smtElem	smoothed element nodal stresses	float	[nel,nen]
strain	principal stress or strain designator	boolean	[1]
sx,y,xy	shear resultants	float	[1]
t	thickness	float	[1]
T	torsion internal to beam or frame	float	[1]
theta	angle change of roof mesh	float	[1]
theP	principal angle	float	[1]
Tinc	increment in angle along mesh edge	float	[1]
Te	element transformation matrices	float	cell
	also used to store B-matrices	float	cell
v	Poisson's ratio	float	[1]
vyz,vxz	shear resultants	float	[1]
Vy,Vz	shear components in beam or frame	float	[1]
wtb	integration weights: plate bending	float	[npb]
wtl	integration weights: elastic lambda	float	[npl]
wts	integration weights	float	[nip]
wtu	integration weights: elastic mu	float	[npu]
wtv	integration weights: plate shear	float	[npv]
x	point coordinate	float	[nsd]
xe	element coordinates	float	[nen,nsd]
xi	integration points	float	[nip,nsd]
xn	nodal coordinates	float	[nnp,nsd]
Xinc	increment along x-axis of mesh	float	[1]
Yinc	increment along y-axis of mesh	float	[1]

References and Further Reading

Bathe, K.J. (1982). *Finite Element Procedures in Engineering Analysis*. Englewood Cliffs, NJ: Prentice-Hall.

Carslaw, H.S. & Jaeger, J.C. (1959). *Conduction of Heat in Solids* (2nd ed.). New York, NY: Oxford University Press.

Hughes, T.J.R. (1987). *The Finite Element Method: Linear Static and Dynamic Finite Element Analysis*. Mineola, NY: Dover Publications.

Love, A.E.H. (1906). *A Treatise on the Mathematical Theory of Elasticity* (2nd ed). Cambridge: University Press.

MacNeal, R.H. & Harder, R.C. (1985). A Proposed Standard Set of Problems to Test Finite Element Accuracy. *Finite Elements in Analysis and Design 1(1)*, 3-20.

MathWorks. (2014). *MATLAB Primer: R2014b*. Natick, MA: The MathWorks.

McGuire, W., Gallagher, R.H., & Ziemian, R.D. (2000). *Matrix Structural Analysis* (2nd ed.). New York, NY: John Wiley & Sons.

Nicholson, W.K. (2003). *Linear Algebra with Applications* (4th ed.). Toronto, ON: McGraw-Hill Ryerson.

Press, W.H., Teukolsky, S.A., Vetterling, W.T., & Flannery, B.P. (2007). *Numerical Recipes: The Art of Scientific Computing* (3rd ed.). New York, NY: Cambridge University Press.

Scordelis, A.C. & Lo, K.S. (1964). Computer Analysis of Cylindrical Shells. *ACI Journal Proceedings, 61(5)*, 539-562.

Timoshenko, S.P. & Goodier, J.N. (1970). *Theory of Elasticity* (3rd ed.). Tokyo: McGraw-Hill.

Timoshenko, S.P. & Woinowsky-Krieger, S. (1959). *Theory of Plates and Shells* (2nd ed.). New York, NY: McGraw-Hill.

Zienkiewicz, O.C., Taylor, R.L., & Zhu, J.Z. (2005). *The Finite Element Method: Its Basis and Fundamentals* (6th ed.). Burlington, MA: Elsevier Butterworth-Heinemann.

Index